CW01021588

Insecurity and Emerging Biotechnology

Brett Edwards

Insecurity and Emerging Biotechnology

Governing Misuse Potential

Brett Edwards
Department of Politics, Languages
and International Studies
University of Bath
Bath, UK

ISBN 978-3-030-02187-0 ISBN 978-3-030-02188-7 (eBook)
https://doi.org/10.1007/978-3-030-02188-7

Library of Congress Control Number: 2018962743

This Palgrave Pivot imprint is published by the registered company Springer Nature Switzerland AG
The registered company address is: Gewerbestrasse 11, 6330 Cham, Switzerland

Acknowledgements

I am grateful to a number of colleagues and friends who have supported different aspects of the underpinning work for this book. This includes my Ph.D. supervisor Dr. Alexander Kelle who fostered my interest in this area. In addition, I have been lucky enough to work alongside a number of supportive friends and colleagues at the University of Bath over the years. There are too many to name, but I am particularly grateful to David Galbreath, Scott Thomas and Timo Kivimaki for comments on earlier iterations of this proposal and the encouragement they have provided. In addition, a number of other friends and colleagues at Bath have also been a constant source of inspiration, not least Mattia Cacciatori, Luke Cahill, Neville Li and Tom Hobson. There is also a wider community of people working on biological and chemical disarmament to whom I am also especially grateful. This includes Brian Balmer, Brian Rappert, Caitriona McLeish, Filippa Lentzos, James Revill, Jean Pascal Zanders, Jo Husbands, Kai Ilchmann, Malcolm Dando, Nick Evans, Piers Millet, Richard Guthrie and Sam Weiss Evans. Naturally, any errors or omissions remain my own.

This work has also been made possible due to a number research and travel grants. This includes a Ph.D. stipend provided by the Wellcome Trust, and post-doctoral work as part of the ESRC-funded Biochemical Security 2030 Project. The latter of these allowed me to develop a deeper appreciation of the complexities of global governance challenges in this area. I would also like to thank Palgrave for ongoing editorial support throughout the process.

Finally, but most importantly, this book would not have been possible without the support of my family.

Contents

Abbreviations

AI	Artificial Intelligence
BTWC	Biological and Toxin Weapons Convention
CCW	Convention on Certain Conventional Weapons
CJVI	Craig. J. Venter Institute
CWC	Chemical Weapons Convention
DARPA	Defense Advanced Research Projects Agency
DOD	United States Department of Defence
DURC	Dual Use Research of Concern
ELSI	Ethical Legal and Social Issues
GMO	Genetically Modified Organism
iGEM	International Genetically Engineered Machine competition
LEAP	Synthetic Biology Leadership Accelerator Programme
MIT	Massachusetts Institute of Technology
NAS	National Academies of Science (US)
NSABB	National Science Advisory Board for Biosecurity
OECD	The Organisation for Economic Co-operation and Development
SAB	OPCW Scientific Advisory Board
Synberc	Synthetic Biology Engineering Research Centre
UNICRI	United Nations Interregional Crime and Justice Research Institute

CHAPTER 1

Introduction

Abstract This chapter introduces the way in which emergent fields of innovation capture the imagination of scientists, policy-makers and publics. It focuses in particular on the rise of proliferation and militarisation concerns about emergent fields of technological innovation. Such concerns are not new; however, today's anxieties reflect contemporary relationships between science, the state and the global order. A central argument of this chapter is that there is a need for new approaches to thinking about the scope, practices and broader politics of governance directed at contemporary techno-scientific fields. To this end, the book argues that we need to understand that policy-making in this area grapples with a number of distinct but interrelated types of problem related to defining the ethical responsibilities of innovators, predicting and managing the societal effects of emergent areas of innovation and managing competitive drives at the international level.

Keywords Disarmament · Innovation · Expertise · New and emerging science and technology

Every so often, a story appears in the popular press which discusses the prospect of a specific scientific breakthrough or technological development being misused by terrorists, criminals or governments. Usually, such concerns exist as a vague anxiety and are accepted as an unwelcome but acceptable consequence of progress or are dismissed as hyperbole or

B. Edwards, *Insecurity and Emerging Biotechnology*,
https://doi.org/10.1007/978-3-030-02188-7_1

misplaced sentimentality. Occasionally, however, innovators appear to take science in a direction which is beyond the pale. This leads to questions about how innovation should be stewarded in the pursuit of some vision of national or global security. It also leads to questions about the moral limits which should be placed upon human inquiry, and the more fundamental relationships between technology and humanity. These questions, or rather contemporary approaches to answering them, are at the centre of this book. This study places a particular emphasis on the challenges raised not only by technologies, but also more fundamentally by the systems we have built to produce them. Importantly, such questions tend to emerge in a tangle—something reflected, for example, in ongoing discussions about the military potentials of Artificial Intelligence (AI).

Recently, we have seen Google programmers (as well as Google's public relations team) grapple with the issue of whether they should work on military projects. In this case, Google had taken a military contract from the US Department of Defence (DOD) as part of Project Maven, a US Intelligence Agency initiative which focused on developing links with the bourgeoning US AI industry. As part of this project, Google was to develop software which could automate some aspects of image analysis. A key challenge that US military intelligence has faced has been sorting through the huge volumes of footage which is collected by drones. As Marine Corps Col. Drew Cukor noted in an update on the project in July 2017:

> 'You don't buy AI like you buy ammunition….There's a deliberate workflow process and what the department has given us with its rapid acquisition authorities is an opportunity for about 36 months to explore what is governmental and [how] best to engage industry [to] advantage the taxpayer and the warfighter, who wants the best algorithms that exist to augment and complement the work he does.'[1]

The announcement of Google's involvement with the project led to disquiet among employees. Despite early reassurances from Google leadership that the project would not develop technologies directly involved in targeted killing, a number of individuals resigned in protest. In April 2017, around 3000 Google employees also signed a letter which called for Google to cancel this contract, and also to pledge publicly not to build warfare technology. In response, Google produced a code of conduct which stated that they would not continue to work with the military on weapons projects directed at people, or which would contravene 'widely accepted principles of international law and human rights.'[2]

This left the DOD with an acquisition problem, which the forces of market competition would undoubtedly solve. It also left programmers working in this area, who were not in principle opposed to the idea of working on weapons projects, in a quandary. Many US technologists may feel that they have a responsibility to help ensure US national security—and to help protect US service people on military operations around the globe. However, even then, it is apparent that this does not give carte blanche in ethical terms—conforming to the letter of the law may not be enough, if they can foresee their work being misused by the state they work for or by others.

Such concerns may have been allayed for some if they had come across an article published in *Nature*[3] as developments at Google unfolded, which argued that scientists needed to continue to work with the US military to further US national and, although more indirectly, international security. The piece noted that the US continued to compete with Russia and China in this area. It also argued that civilian work would potentially end up being exploited even if US programmers refused to work on government defence projects—the only difference would be that it would enter weapons technology through more indirect routes. The author of the piece noted that the developers of mass-produced remote-controlled drones, would not have envisioned the technology ending up in military applications or foreseen the way in which this technology would be hacked in battlefields around the world.

In the context of the inevitable exploitation and proliferation of technology, it was argued that the US AI community should continue to work with the US military, and encourage the government to exploit the technologies they were developing in an 'ethical' way. According to the piece then, it was up to developers to make up their own minds about whom they worked with, and on what projects, on a case-by-case basis, as the article noted:

> Some proposals will be unethical. Some will be stupid. Some will be both. Where they see such proposals, researchers should oppose them.

This then placed the innovator and their apprehensions of the world at the centre of how the challenge posed by the omnipotence of technology was to be understood. Google would go on to claim in a publicly shared memo that they would not invest in:

> Technologies that cause or are likely to cause overall harm. Where there is a material risk of harm, we will proceed only where we believe that the benefits substantially outweigh the risks, and will incorporate appropriate safety constraints.[4]

Why this is an admirable aspiration—it remains just that. Many in society are still in disagreement about the social effects of inventions even with the benefit of hindsight; from the motor car, to plastics, and nuclear weapons. With emergent technology comes even greater uncertainty and ambiguity. Framing the issue in terms of the potential apprehensions and associated ethical responsibilities of single inventors, research groups or corporations appears to arbitrarily narrow assessments to debates about what we can reasonably expect innovators to foresee, and their judgement of the potential benefits and harms. It also potentially contributes to the uncritical acceptance of broader structures of investment and oversight.

Despite this, our eye is often drawn to innovators at the cutting-edge when thinking about the problem of technology control. This is in part perhaps because of a broader philosophical disposition towards individualism, particularly in Western ethical thought. It also reflects broader cultural norms about innovators and the relationship they have with their creations, as well as norms relating to the allocation of proprietary rights to inventors which have increasingly come to dominate globally. It is also because scientists tend to be early advocates of emergent technology, and occasionally, key proponents for control.

Indeed, the biographies of innovators have taken on lives of their own—and have become heroic epics, as well as parables about the moral weight of discoveries upon inventors, hubris and at times greed. This has been acutely true for those involved in weapon development—from the machine gun to the Manhattan Project. At the same time, we have also seen scientists lead drives for restraint and prohibition. This includes scientists who sought to ban the bomb, scientists who campaigned for comprehensive biological and chemical weapon disarmament, as well as those who campaigned against Agent Orange and other environmental weapons. Today, technologists considering the ethics of the weaponisation of artificial intelligence are grappling with the same dilemmas that their predecessors did—mediated as they are by the time and place they find themselves in.

Indeed, briefly returning to the issue of military research at Google, a curious example of an act of protest would emerge. One resigning employee began a petition to rename a Google conference room after Dr Clara Immerwahr. Clara Immerwahr was a twentieth-century chemist who was married to Fritz Haber, famous for his Nobel Prize winning work on the industrial production of ammonia. Immerwahr committed suicide following an argument with her husband about his role in the German chemical warfare effort—shooting herself with Haber's military revolver. This story has been subject to several retellings over the years,[5] but as far as events at Google are concerned, we can be pretty clear on the type of point that reference to this tale was meant to make.

There is no denying that many scientists have risked and lost much in the name of principle. There is also a certain romance in these acts of heroism, and the Promethean dread and Faustian pacts of innovators.[6] And this way of thinking and writing about these problems is endemic. In the field of virology, for example, there have been concerns about certain lines of work on viruses with global pandemic potential such as avian influenza. These focus particularly on work which seeks to create laboratory versions of disease which are more deadly, and more difficult to treat—in order to stay ahead of natural evolutionary purposes, and develop a deeper understanding of underlying mechanisms of pathogenicity.[7] This had led to concerns framed in terms of public safety and proliferation risks, as well as questions about whether such work blurs the line between peaceful and offensive research. The work of specific scientists has become a flash-point of broader ongoing debates on the ethics of such work—and questions about the ethical responsibilities of scientists have been placed front and centre.[8]

This means that it is particularly easy to forget that questions of ethics often extend far beyond the agency of individual innovators and the communities they work in. The innovator's dilemma is embedded in broader questions about the appropriate role and goals of innovation within societies. This includes the need to balance precaution with militaristic, economic as well as exploratory drives. The way in which these questions are apprehended, and balances sought, varies between innovation communities and national contexts.

Such questions are also opened up much more readily in new fields of innovation than established ones. In part, this openness seems to stem from the sense of novelty associated with emergent technologies. This novelty has two dimensions. On the one hand, the breathless promissory discussion of emergent fields emphasises the powerful transformative potentials contained within them. Emergent fields are framed as having the potential to revolutionise every aspect of our lives. On the other hand, this emphasis on boundless potentials reasserts the impossibility of predicting and preventing negative aspects of a technological field. A second dimension of this apparent openness seems to stem from the fact that in the early years of a techno-scientific field, its organisation and goals, as well as the potentials of specific technologies, are more open to broader societal debate and discussion than established fields. The myth of 'pure science', which still persists in many national contexts, often elicits the idea that emergent fields represent virgin political territories which have yet to be colonised and corrupted by politics. At other times, it is the more latent politics in innovation which seem to be awakened. The announcement of a new 'Space Force' by President Trump has opened up old debates about the role of the institution of science in the service of national security—and not just the ethical questions facing scientists. Such debates do not necessarily signify moments of political change, but they do signify a potential site of it. In this context, societies attempt to navigate a path between paranoia and hubris, and ensure that the technologies they foster serve some conception of the public good, provide the resources that societies demand, while still remaining compatible with the values they hold and project internationally. However, these systems of management are under constant challenge. Our regulatory and moral institutions appear to struggle to keep pace with the changing distribution and economics of innovation, associated transformations in the character of organised violence, and the evolving structure of the international system.

Improvised explosive devices[9] and drone technology make for excellent case studies for understanding the way in which conflicts function as both consumers and producers of innovation—and in particular the massively parallel social processes through which emergent technologies become integrated into and transform organized violence. Such routes may often be side-on and emerge from improvisation on the ground. An example is the increasing use of social media and civilian software programs in warzones by combatants, in an ever-increasing range of ingenious ways. In 2016, for example, it was widely reported that Ukrainian military artillery units had fallen victim to a new form of hack.

An Android application had reportedly been developed and distributed by a Ukrainian military officer to make artillery targeting easier for troops in the field. The software was promoted to Ukrainian service personnel via an online forum—access to the software was controlled by the developer. At some point, however, the pro-Russian 'Fancy Bear' cyber espionage group circulated a 'hacked' version of this application, which would be shared via the same online forums. It is claimed that this may have allowed Russian forces to track and destroy Ukrainian artillery.[10]

Not only have societies become increasingly aware of technology as a driver of change in national and international security, but they have also worried more about it. Especially in relation to the impacts of new technology on the norms of warfare, and the barriers to mass casualty terrorism. Such anxieties have tended to manifest as challenges for experts,[11] as well as for broader systems of expertise. This relates to the more general problem facing contemporary political systems. Modern societies have developed a preference for scientific forms of decision-making, while at the same time place unreasonable demands upon these scientific forms of knowledge. This is in terms of both requests to 'extend' the scope of scientific dominion to non-scientific problems, and demands for decisions to be made more quickly than a genuine scientific consensus can be reached. This generates pressures on individual technical experts, expert communities, and institutionalised expert bodies. This in turn creates demands for supplementary forms of expertise both within and outside of these expert communities.[12]

On another level, however, these are also problems of *purpose*. Gazing into technological futures, and arguments about the societal impacts of emergent technologies, also form part of deeper introspective quests to define and redefine the relationship between societies and the broader project of scientific and technological progress. This includes the role of scientific and technological visions of how and why states fight wars, and how states maintain domestic order. While this second set of questions sounds more abstract, these issues are ever present, explicitly or implicitly, in the evaluation of the security implications of emerging technology—domestically, and increasingly at the global level. The ongoing internationalisation of innovation will undoubtedly also continue to reveal that assumptions made in the West about the deeper social and political purposes of technology are just that. This will mean that the challenge of building global institutions will not be solved purely by resolving disagreements on the most effective means to exploit technology. It will also require attention to deeper questions about the relationship between conceptions of technology, innovation and the political. Indeed, these are not new questions, and have been touched upon as part of the broader projects to develop global norms around sharing the cultural and economic benefits of science. This includes, for example, work under the auspices of the Universal Declaration of the Human Rights and the International Covenant on Economic, Social and Cultural Rights.[13] Such questions are ever present in debates about new technology—and yet often appear to be absent in the instruments which are applied in the governance of these technologies.

The Central Contribution of This Book

The purpose of this book is to argue that emergent technologies are a site of political domination, resistance, contestation,[14] and of 'criticality' in all its various forms.[15] Technology is not only used as a tool to reproduce existing institutions and power relations; it becomes a site at which they are challenged, and this complicates quests to control. This also means that we often speak at cross-purposes when considering control of new and emergent technological developments; and this is acutely true for matters of national and international security. I argue in this book that the starting point in making sense of the scope, practice and politics of this area is to realise that they revolve around three central. By Paradoxes, I simply mean three relatively stable and commonly identifiable sets of predicament which manifest as a challenge facing individuals, organisations as well as governments. These are the innovator's paradox, the innovation paradox and the global insecurity paradox Box 1.1. While some aspects of these predicaments are ancient, much that characterises their contemporary manifestation are much younger. To give an example, worries about the theft of military technologies are certainly older than either the words 'military' or 'technology.' Today, however, the way we think about these problems is heavily informed by developments in the past 100 or so years.

Each paradox appears to generate contradictory demands on individuals, states and the international community. These paradoxes occasionally emerge as more acute ethical dilemmas facing specific actors or communities but more often they are resolved more unconsciously and implicitly through habit and appeal to tradition. Working through the more essential and contingent aspects of each of these problem framings highlights different dimensions of the challenge of making misuse potentials of technology governable.

Box 1.1 The Three Paradoxes of Innovation Security

The innovator's paradox: Innovation can produce both good and negative consequences. This then appears to generate conflicting ethical responsibilities for those that create, and those that facilitate creation.

The innovation governance paradox: Societies seek security through the development and maintenance of innovation systems, but innovation can also generate insecurity. This then appears to create conflicting demands for exploitation and precaution.

The global insecurity paradox: The route to national security is more global approaches management of technology. But national security is at the centre of current approaches to global security.

In this book, I focus in particular on the extent to, and ways in which, these paradoxes have manifested in emergent publicly funded techno-scientific projects which are primarily of a civilian nature. I argue that these fields are sites at which long-standing ideational and power struggles play out in the manufacture of ethical principles structured by prevailing political institutions. However, they are also spaces in which new ways of governing innovation, and visions of innovation, are imagined, experimented with and co-opted into the existing landscape. As with all areas of human affairs, attempts to govern emergent technology are relient on three forms of collective practice: ethical evaluation (i.e. establishing what dominant social conventions require), policy design (i.e. working to achieve these 'ethical' outcomes) and political decision-making (i.e. in as far as reasserting or transforming the scope of what is ethical and what is practical). While it is possible to make *analytical* distinctions between these dimensions—they are actually tightly integrated. This means that the very same initiatives can take on different types of significance, and have different forms of cross-cutting implication. It also means accepting that debates about specific technologies and fields of innovation are embedded in much broader political struggles to define and redefine the limits of violence, the purpose of scientific innovation, as well as dominant modes of governance within societies.

A second key contention in this work, stemming from engagement with security studies, is that that rhetorics, practices and politics of national and international security play obvious as well as more subtle roles in this area of human affairs.[16] This is reflected not only in those institutions which serve overt security functions, but also in deeper entrenched structures which constitute the relationship between innovation, the state as well as national societies—including those which relate to inter-state competition. This requires us then to take seriously the idea that security concerns can, to some extent, be *analytically* distinguished from broader societal apprehensions about emergent technology; but that they also remain tightly intertwined in the broader systems of innovation and governance which give rise to them.

Book Focus, Method, Study Design and Structure

My central aim in writing this book is to develop these critical insights in a way which will highlight existing cultural and institutional limits and possibilities for change in this policy space. The book draws upon a long tradition of such work in the field of disarmament. It focuses on an issue which has always been acknowledged in this field, but which has until relatively recently been on the periphery. In recent years however, pre-emptive and precautionary approaches to innovation governance have become increasingly important in this area of study. It is hoped that this book helps to reveal the complex and contingent character of the security politics which surrounds emergent techno-scientific fields. In this respect, the work has benefitted in particular from engagement with more cultural approaches to studying the role of technology in national security and warfare,[17] and the culturally specific character of technology governance more generally.[18]

In methodological terms, this book draws upon a critical historical study[19] in which central concepts, themes and arguments have been iteratively refined with the goal of developing policy-relevant insights over several years—a process which has benefitted from feedback not only from scholars but also from practitioners associated technology assessment, security and disarmament. The work synthesizes empirical research conducted by the author; conducted as part of a number of projects. This includes an initial comparative historical study which compared UK and US approaches to addressing security aspects of the field of synthetic biology, which drew upon policy documents, semi-structured interviews, participant observation and ongoing correspondence with scientists and other policy shapers in this space. This initial fieldwork would be supplemented by my involvement with an initiative which brought together academics and policy shapers to develop recommendations for how to improve the formal science and technology and review processes in the context of the BTWC (Biological and Toxin Weapon Convention).[20] Further to this, I worked for two years on a project designed to develop new understandings of how to improve responsiveness to the global chemical and biological weapon regimes to advances in science and technology.[21] This latter project would involve the organisation of a number of workshops, conferences and other events in Bath, London and Geneva. During these projects, I had the opportunity to speak to, observe and collaborate with many experts and practitioners who would be a source of ideas as well as critique.

The work speaks to the practical challenges of governing techno-scientific fields, but also to some of the deeper political and philosophical questions that developments in this policy space have raised. In this respect, this work may be of interest to those working on humanitarian issues around new technologies, scientists thinking about the ethical implications of their work, as well as those interested in the study of the governance of technology and security issues more generally. I ask that experts and practitioners in the case-study field I examine forgive omissions on more technical and procedural issues—these are of course inevitable in this form of presentation.

The book is structured as follows. In the second chapter, I unpack the three paradox of innovation security. This is followed by an introduction to the techno-scientific field of synthetic biology as a site of security politics in Chapter 3. Chapters 4–6 involve a more in-depth presentation of the development of this field, associated security concerns and governance responses, as well as an overview of the specific types of challenges and dilemmas raised by techno-scientific fields such as Synthetic Biology. This is followed by a concluding chapter, which summarises key findings in relation to the case-study field, and makes a number of suggestions for future work.

Notes

1. Cheryl Pellerin, 'Project Maven to Deploy Computer Algorithms to War Zone by Year's End', U.S. Department of Defense, 21 July 2017, https://www.defense.gov/News/Article/Article/1254719/project-maven-to-deploy-computer-algorithms-to-war-zone-by-years-end/.

2. Sundar Pichai, 'AI at Google: Our Principles', Google (blog), 7 June 2018, https://www.blog.google/technology/ai/ai-principles/.
3. Gregory C. Allen, 'AI Researchers Should Help with Some Military Work', News, Nature, 6 June 2018, https://doi.org/10.1038/d41586-018-05364-x.
4. Sundar Pichai, 'AI at Google: Our Principles', 7 June 2018, https://www.blog.google/technology/ai/aiprinciples/.
5. Bretislav Friedrich and Dieter Hoffmann, 'Clara Immerwahr: A Life in the Shadow of Fritz Haber', in *One Hundred Years of Chemical Warfare: Research, Deployment, Consequences* (Cham: Springer, 2017), 45–67, https://doi.org/10.1007/978-3-319-51664-6_4.
6. See for example R. W. Reid, *Tongues of Conscience: War and the Scientist's Dilemma* (London: Constable & Company Limited, 1969).
7. For a recent review see Michael J. Imperiale, Don Howard, and Arturo Casadevall, 'The Silver Lining in Gain-of-Function Experiments with Pathogens of Pandemic Potential', in *Influenza Virus: Methods and Protocols*, ed. Yohei Yamauchi, Methods in Molecular Biology (New York, NY: Springer, 2018), 575–87, https://doi.org/10.1007/978-1-4939-8678-1_28.
8. Brett Edwards, James Revill, and Louise Bezuidenhout, 'From Cases to Capacity? A Critical Reflection on the Role of "ethical Dilemmas" in the Development of Dual-Use Governance', *Science and Engineering Ethics* 20, no. 2 (June 2014): 571–82, https://doi.org/10.1007/s11948-013-9450-7.
9. James Revill, *Improvised Explosive Devices: The Paradigmatic Weapon of New Wars* (Cham: Springer, 2016).
10. Dustin Volz, 'Russian Hackers Tracked Ukrainian Artillery Units Using Android…', *Reuters*, 22 December 2016, https://www.reuters.com/article/us-cyber-ukraine/russian-hackers-tracked-ukrainian-artillery-units-using-android-implant-report-idUSKBN14B0CU.
11. Understood here as individuals, communities and institutions embodying a delimited range of specialist knowledge and skills which are acknowledged within the broader political system.
12. See, for example, Harry Collins and Robert Evans, *Rethinking Expertise* (Chicago and London: University of Chicago Press, 2009).
13. Aurora Plomer, *Patents, Human Rights and Access to Science* (Cheltenham: Edward Elgar, 2015).
14. Antje Wiener, *A Theory of Contestation*, 2014 edition (New York: Springer, 2014).
15. Andrew Feenberg, *Critical Theory of Technology* (New York: Oxford University Press, 1991); Darrell P. Arnold and Andreas Michel, *Critical Theory and the Thought of Andrew Feenberg* (Cham: Springer, 2017).

16. Thierry Balzacq, 'Enquiries into Methods: A New Framework for Securitization Analysis', in *Securitization Theory: How Security Problems Emerge and Dissolve*, ed. Thierry Balzacq and J. Peter Burgess (Abingdon: Taylor & Francis, 2010).
17. Especially Christopher Coker, *Future War* (Wiley, 2015); Andrew Cockburn, *Kill Chain: The Rise of the High-Tech Assassins* (New York: Henry Holt and Company, 2015).
18. Stephen Hilgartner, Clark Miller, and Rob Hagendijk, *Science and Democracy: Making Knowledge and Making Power in the Biosciences and Beyond* (New York: Routledge, 2015); Sheila Jasanoff, *Designs on Nature: Science and Democracy in Europe and the United States* (Oxfordshire: Princeton University Press, 2005); Sheila Jasanoff and Sang-Hyun Kim, *Dreamscapes of Modernity: Sociotechnical Imaginaries and the Fabrication of Power* (Chicago and London: University of Chicago Press, 2015).
19. Martin Reisigl, *The Discourse-Historical Approach* (Routledge Handbooks Online, 2017), https://doi.org/10.4324/9781315739342.ch3.
20. Alexander Kelle, Malcolm R. Dando, and Kathryn Nixdorff, 'S&T in the Third BWC Inter-Sessional Process: Conceptual Considerations and the 2012 ISP Meetings' (Bradford Disarmament Research Unit, University of Bradford, 2013), http://www.brad.ac.uk/acad/sbtwc/ST_Reports/ST_Reports.htm.
21. See, www.biochemsec2030.org.

CHAPTER 2

The Three Paradoxes of Innovation Security

Abstract This chapter outlines some of the central concepts and approaches to governance which dominate discussions of misuse concerns associated with emergent fields of innovation. The key conceptual contribution of this book is to highlight that we often speak at cross-purposes when considering control of new and emergent technological developments, and that this is acutely true for matters of national and international security. A good starting point in making sense of this is to realise that politics in this space has come to revolve around three central paradoxes which generate a range of apparent dilemmas for innovators and societies more broadly. While aspects of these dilemmas are ancient, much that characterises their contemporary manifestation are far younger than most people would assume. In this chapter, each of these paradoxes is introduced in turn.

Keywords Responsible research and innovation · Technology assessment · Disarmament

The Innovator's Paradox

Humans are by nature curious and inventive. Technological innovation reflects the fostering and exploitation of this tendency at societal level—through investment into individuals, communities and institutions. Technologists, motivated by the technical appreciation they have of

B. Edwards, *Insecurity and Emerging Biotechnology*,
https://doi.org/10.1007/978-3-030-02188-7_2

developments, as well as the moral sense of responsibility for their creations, have often reflected on their ethical responsibilities. In this book, there are two questions of particular importance. First, which forms of innovation, by virtue of their character, products or intended use, should be off limits? And second, what responsibilities do scientists have to attempt to prevent the use of their work by others in ways that would contravene these limits?[1]

The answers to these questions tend to be present explicitly or more implicitly in law and codes of conduct. But more fundamentally, they are reflected in the more nebulous, social and professional conventions which have emerged within innovation communities. These norms delineate the community's relationship with broader society and the state. These vary between time and place. They also continue to evolve, as part of self-conscious attempts within communities to articulate and redefine ethical standards of behaviour.

Generally speaking, most contemporary discussions of the responsibilities of innovators speak to two prevailing ideas, which are commonly juxtaposed—both are essentially mythology, but do seem to reflect the inherent tensions between 'doing' and 'reflecting' upon the meaning and implications of innovation. On the one hand, you have what are commonly referred to as 'pure' visions of innovation, which can be traced back through the history of the emergence of the scientific method. Tom Lehrer, who grew famous for his line in satirical songs, picked up on this when he wrote about the German scientist Wernher von Braun who had built rockets for the Nazi regime. He was one of many scientists who would be recruited into military and civilian programmes in Western states at the end of the Second World War. In 1965, Lehrer sang:

> 'Don't say that he's hypocritical. Say rather that he's apolitical "Once the rockets are up, who cares where they come down? That's not my department" says Wernher von Braun'.

This caricature draws upon an ethos of scientific innovation which had come to dominate by the twentieth century, and is still reflected today to varying extent within different scientific communities—and these sentiments are certainly recognisable by most practising scientists. This idea of 'pure' science is characterised by an emphasis on the collective principles which provide assurances against the corruption of the field by personal and other special interests and to ensure that scientific developments are shared.[2] This vision then tends to emphasise the internal characteristics of the scientific establishment which enables the development of 'good' science—as opposed to making ethical assertions on the broader social purposes and implications of innovation. It is an instrumental vision of innovation, but one which externalises consideration of the purposes to which the instrument is put. On the other hand, contemporary science, which creates a significant demand for resources, has become increasingly integrated into the societal and political fabric of states. This is reflected in the politics which has surrounded the major national and international 'big' scientific projects which have emerged since the end of the Second World War. These projects have emphasised different stances on the political and moral expectations which societies should place upon science, as well as different visions about if and how scientific innovation should be integrated with industry and military institutions. These two types of ethos then, which emphasise the idea of 'good science' and innovators as 'good citizens', have fundamental impacts on the ethical apprehensions of individuals; they also vary widely between different communities of innovators and in comparable fields in different national contexts. It is also clear that a comparable tension exists in business ethics—reflected in contemporary drives to develop corporate responsibility.[3]

At the same time, the role of innovators has always gone beyond following ethical conventions, in that it has also involved shaping these conventions. Technological innovators have historically played a number of types of role which have impacted on the relationship between specific fields and the security apparatus, as well as the broader social contract between innovation and the state. Scientists, then, have operated in a wide range of ethical contexts, and have both resisted and encouraged the weaponisation of science.

On the one hand, we have seen advocates of technology attempt to resist limits that have been placed on new technology by squeamish publics. This has included advocates of military and weapons technology—who often point not only to the value of development and adoption of new technology based on national security arguments, but also humanitarian ones. Recently, for example, advocates of the expansion of US efforts in the area of lethal autonomous weapon systems have pointed to the idea that the failure to do so will harm US security in the long term, as adoption by other states is inevitable.[4] They also argue that such weapon systems may pose reduced risks to non-combatants as compared to existing weapon systems. However, optimism about the ability of new and foreseen weapon systems to transform the inherent brutality of warfare, and in particular to reduce human suffering, has, of course, not always been well placed. This is in part because it is difficult to predict the operational cultures which will emerge around new weapon systems—something revealed particularly well, for example, in work on safety[5] and doctrinal aspects[6] of nuclear weapons, as well as smart-bombs, drones and fully autonomous weapon systems.[7] It is also because such assessments tend to be written by those who will sit far behind the weapon sights rather than directly in front of them. Richard Gatling, the inventor and name-sake of an early type of machine gun, reportedly hoped that such weapons could reduce the human cost of war.

There are comparable examples of hubris in the field of chemical warfare. Following the end of the First World War, many state militaries were unsure if chemical weapons were a weapon of the future worth investing in, or if utility was limited to fighting a type of warfare which the world would never see again—gas masks were apparently abandoned *on en masse* by the US military as they departed Europe.[8] J. B. S. Haldane, a British scientist writing in 1925, voiced a perspective which was held by many within chemical weapon research and development programmes at the end of the First World War. Written at a time when chemical weapon research and development institutions were fighting to secure resources, and civilian chemical industries were still in recovery,[9] Haldane dismissed the prospect of a comprehensive prohibition of chemical weapons as sentimental. He argued that the ethical objection to 'scientific weapons' of the late First World War, including choking and blister causing agents, were essentially the product of ignorance and fear of the new.[10] Such technological optimism came despite the fact that he himself foresaw the aerial bombardment of cities with chemical weapons, and saw great potentials in the ability to increase the lethality of such weapons. Indeed, sulphur mustard was used by the Spanish in the Rif War (1921–1927),[11] which is thought to be the first example of the agent being delivered by air. This would be followed by the large-scale and systematic use of the agent by Italy in Abyssinia. Italy had technological superiority, including a modern air force. As one history notes, '*The use of sulphur mustard was particularly effective because the Ethiopian soldiers wore traditional light desert garb that exposed the skin. In addition, Ethiopian soldiers typically wore sandals or were bare foot.*'[12] Largely unopposed in the air, they even experimented with spraying the agent by the gallon from the back of a planes. An estimate from the period put the casualty figures in the tens of thousands.[13] Several types of chemical agent, including sulphur mustard also have a notorious level of open-air environmental persistence and remain highly toxic for decades in containers.

Indeed, mustard shells which are remnant from use in Europe in the First World War, and which were also stockpiled in the Second World War, still occasionally cause injuries in Europe.[14] In addition, China is still dealing with the problem of stockpiles of chemical weapons abandoned on its territory over 70 years later. There have been well over 2000 casualties (including some deaths) caused by abandoned munitions in China, with the most recent fatality reportedly occurring in 2003.[15]

Haldane also underestimated the destructive potential that would eventually be leveraged out of organic chemistry in the development of new powerful nerve agents and environmental weapons. Haldane reasoned in 1937 that[16]:

> We have probably not completed the list of possible poisonous gases or rather, let us say, poisonous volatile compounds, because many of the so-called gases are liquids at ordinary temperatures. Nevertheless, I do not believe in the probability of anything very much worse than mustard gas.

Over the next twenty years, several states developed new types of chemical agents. This included the G-series of nerve agents (tabun, sarin, soman and then cyclo sarin) as part of the German chemical weapon programme. Tabun would be discovered by the military only as a result of hunts for candidate pesticides within the civilian chemical industry. This was followed by the development of the V-series of agents (including most notably VX, which was developed by the UK in the 1950s). As with Tabun, the discovery of this series of nerve agents can also be traced directly to an agent which was incidentally discovered as part of civilian work on pesticide development. Indeed, one of the less lethal agents in this series (VG) would actually be marketed as a pesticide called Amiton.[17] However, by the end of the 1950s, military establishments were taking a much more active approach in the identification of candidate agents. VX would be discovered, for example, at Porton Down in the UK in the late 1950s. The UK, as well as a number of other state militaries, would go on to identify and develop the wide range of both lethal and less lethal agents which are recognised as chemical warfare agents today.

Haldane also underestimated the persistent effects of chemical weapons and degradation products in the environment when used at industrial scales:

> 'I have stated it is unlikely that anything much worse in that line will be made. Therefore, when one reads in H.G Wells' The Shape of Things to Come about chemical compounds which will render large districts uninhabitable for many years, one pays a tribute to his imagination, but not so great a tribute to his knowledge of organic chemistry.'

During the Vietnam War, the US would employ chemical agents, including Agent Orange, as part of its environmental warfare campaign. Many of the most heavily targeted areas would indeed remain habited after the war—but these weapons would continue to kill and injure, causing birth-defects and cancers. Environmental contamination still persists over forty years later.[18]

Haldane was also equally dismissive of the potentials of biological warfare in the 1930s, citing the challenges of engineering and weaponising such agents. And yet, biological weapons were employed by Japan in the Second World War,[19] and states went on to develop a wide range of biological weapon systems involving a large number of agents in the post-war years. This included plague, anthrax, tularemia and smallpox—as well as a host of other plant- and animal-based pathogens,[20] although most of these systems would never see use. This was due at least in part to the threat of in kind, or even nuclear retaliation; it was also based on the increasing realisation that great-power monopoly over advanced biological weapon systems would be short-lived, and that proliferation was inevitable. Indeed, a recent review suggests that by 1990, as many as eight states had active programmes.[21]

Haldane saw no hope of a prohibition for chemical or biological weapons, and so saw little point in resisting processes of militarisation. This was, in part, because he saw the endless hostile exploitation of every area of science as inevitable. Other scientists also put forward humanitarian and deterrent-based arguments for the development of biological weapons throughout much of the twentieth century.[22] And yet, today, both biological and chemical weapons are almost universally recognised as abhorrent weapons of war, and both are comprehensively prohibited under international law. The emergence of norms against their development and use has coincided with the gradual throttling of an area of military exploitation which would surely have made the world a much more dangerous place. This is a belief which motivated a number of former weaponeers and other scientists to campaign against the development of these weapon systems in the early years of the global prohibition regime and has inspired continued investment into these systems in more recent decades.[23] This history also resonates in debates about the military potential of emergent fields of biotechnology and advanced technology more generally. In the field of autonomous weapons, for example, it is reflected in debates about whether the unrestricted hostile exploitation of technological advance in the development of weapon systems is inevitable, whether technology can somehow purify war, whether these technologies could fall into the 'wrong' hands, and whether certain types of systems should be allowed to taste blood before we pass judgement on net humanitarian benefits.[24]

We have inherited a world which bears the scars of attempts to both open up and close down the destructive potential of different areas of technology. And scientific ethics and expert activism are central to this history. Scientists have been advocates of military exploitation and have also been central to the emergence and continued evolution of contemporary global weapon prohibitions and International Humanitarian Law. They have played fundamental roles as campaigners in the early days of many regimes, and as a continued source of technical expertise throughout the lifespan of treaty systems—in the fields of biological, environmental, space, chemical and nuclear disarmament.

It is clear, then, that scientists are often important in the apprehension of the potentials of technology, and often exhibit a significant degree of foresight. They also play a wide range of supporting roles in the governance of technology. Examining the perspectives of scientists also helps us to get to grips with the ethical context which scientists operate in as well as the agency they have. However, to focus on the apprehensions of innovators exclusively potentially leads us to ignore the wider relationship between the broader institutions of innovation and national and international security—to the neglect of deeper but no less practical questions about the need to update the laws of war in the context of ongoing shifts in the configurations between broader institutions of innovation, conflict and society more broadly.

The Innovation Paradox

The innovation paradox centres on the apprehension that technology can have a wide range of both positive and negative impacts of society—especially in terms of national and international security, and that attempts should be made to maximise the positive effects, and minimise the negative. A key challenge societies face, however, is not only predicting the impacts of specific technologies, but also identifying and dealing with the negative effects once they are legion. In the previous section, it was noted that new, more 'applied' visions of innovation have come to dominate within the sciences, and that this has shaped the ethical context in which innovators operate. In this section, we get to grips with these broader shifts more concretely—and in particular, the implications which have flowed from changing approaches to innovation at the national level, the emergence of broader societal apprehensions about innovation, and the emergence of new approaches to governing these concerns.

Particularly since the end of the Second World War, exchanges between military and civilian science have been actively encouraged—and technologies, as well as the systems of innovation which give rise to them, have become increasingly integrated. However, this integration raises challenges for the control of technology from a security perspective. This includes the prevention of the proliferation of 'taboo' weapons, which are weapons against which there are stigmas regarding development and use. In the context of nuclear, chemical and biological arms control, there are technologies and materials which are necessary for both maleficent and benign applications. For example, uranium can be enriched in order to produce nuclear power, as well as nuclear weapons. In this context, the term 'dual-use' refers to technologies that are understood to have both legitimate peaceful (civilian) and illegitimate or controlled (military) applications within the international community. Since at least the end of the Second World War, states have attempted to systematically control who has access to dual-use technologies, primarily through systems of licensing and the harmonisation of export controls. Up until around the turn of the twenty first century, much less consideration was given by states to controlling civilian research and emerging technology because of the potential for hostile misuse.

In the second part of the twentieth century, however, several areas of development contributed to the emergence of more sustained attempts to both harness the military potential of technologies as well as control proliferation. First, this has included ongoing reconfigurations between military and civilian innovation. In the post-war years, military innovation was driven primarily by large state run programmes—something which suited containment strategies to mitigating the theft of technologies by other states—as seen in the field of nuclear technology, for example. However, as civilian and military innovation has become increasingly integrated, technologies, as well as their means of production, became increasingly diffused. This is also coupled with the rise of increasingly neo-liberal approaches to investment and governance—something which has made it harder for states to both control and track security-relevant developments.

At the same time, this increasingly decentralised model of innovation has also been coupled with the rise of more ambitious state-level planning in the areas of both innovation strategy and technology assessment. This then has generated an increasing tension between drives to foster and exploit innovation through harnessing market forces, and the desire to control innovation in terms of its outputs and societal implications. This tension has given rise to a dilemma which was most famously described by David Collingridge in the 1980s as a double-bind problem that:

> 'The social consequences of a technology cannot be predicted early in the life of the technology. By the time undesirable consequences are discovered, however, the technology is often so much part of the whole economics and social fabric that its control is extremely difficult. This is the dilemma of control. When change is easy, the need for it cannot be foreseen; when the need for change is apparent, change has become expensive, difficult and time consuming'.[25]

Collingridge was writing at a time at which emphasis was placed on the idea that national-level science planning could rationalise the innovation process—and that the state could rapidly increase rates of innovation through technocratic approaches to governance. In this context, linear conceptions of innovation would emerge—in which technologies were assumed to progress from basic scientific knowledge through a development process which leads ultimately to the diffusion of the innovation and its eventual consequences. Within this understanding, different forms of social intervention can take sequentially along this pathway—something which made sense at a time in which there were concerns that the technical and social potentials emanating from science were being stove piped into technocratic systems of technology development in both the East and the West.[26]

However, broader transformations in the character of innovation, which can be traced through the emergence of more neo-liberal and market driven approaches, have meant that linear models have given way to more systemic visions of the innovation process. In this context, innovation is understood as the product of an increasingly integrated matrix formed between university, industry and the state which centre on sites of '*co-production*'[27] which integrate processes of scientific research, commercialisation and societal evaluation. This relationship is managed in different ways in different national contexts, which is reflected in the distinct constellations of expert and regulatory bodies directed at the governance of emerging technology.[28] Generally speaking, however, these broader transformations have been associated with the increasing integration of processes directed at anticipating and deliberating upon the societal implications of emergent fields, as well as on the development of more proactive approaches to public engagement as well as ethical and risk assessment within research institutions. In recent decades, concerns about terrorism, as well as military investment into new fields (and in particular, the cyber and biological domains under the auspices of the US Third Offset Strategy)[29] have meant that the discussion of security issues has become an increasingly important aspect of ethical debates about a wide range of areas of innovation.

In this book, a particular focus is upon attempts to harness technology in the name of national and international security, governing emergent technologies through existing systems of control, as well as in the development of new modes of control. Each of these areas has been associated with distinct political challenges, which vary between national contexts and fields. In relation to exploiting technology, it is clear that these questions of military technology development and adoption are shaped by military culture and broader societal norms impacting upon questions of which technologies should be exploited, to what ends, as well as the political processes and ethical criteria which should guide the military exploitation of technology—including in the area of weapon development. This is often discussed in terms of the competing drives to annihilate enemies and for moral restraint that are embedded in the norms of warfare. The increasing centrality of science and technology to states' understanding and practice of war, as well as the criteria by which technology are assessed, reflects both of these drives.[30]

The challenges raised by emergent technology relates then to *managing* the negative consequences of emergent technology. In this area, this essentially boils down to limiting potential secondary negative impacts on security (e.g. proliferation) and unintended humanitarian consequences of such work. Increasingly, such management has been parsed in terms of the management of complex or *systemic risk*.[31] Systemic risks can be defined in contrast to 'simple' risk problems, which generally involve the unquestioned application of predefined institutional decision-making routines to problems as they manifest. In comparison, systemic risk governance problems are characterised by complexity, uncertainty and ambiguity. These issues are complex, as they do not involve simple causal chains of events with easily quantifiable consequences, but rather a large set of intervening variables. These issues are uncertain, as there are insufficient data or information to convincingly assess the probability and outcomes of bad events. They are also ambiguous, as they typically include conflicts over values, such as different stakeholders taking contrasting 'legitimate' standpoints on a given issue.

To give an example, the risks posed by bioterrorist scenarios are complex because of the number of variables in threat assessment, with regard to the technology as well as to the intentions and capabilities of actors. There is also an absence of criteria for measuring the probability and effects of misuse of research and technology by terrorist groups, meaning discussions are largely based on analogy.[32] Finally, there is ambiguity with regard to how the values of 'scientific freedom' and 'security' should be conceptualised and balanced. Systemic risk governance has several key dimensions; of particular interest in this work is the 'pre-assessment' phase. The term 'pre-assessment' refers to the normal political process by which novel issues (i.e. risks/threats) are initially identified and constructed as governable problems within modern societies. Renn has identified four interrelated components that would be expected in the pre-assessment stage or in an emerging systemic risk regime.[33] The first component is the framing of the problem, which often involves disagreements over problem definition (such as the scope, severity and causation). The second component involves systematic searches for new hazards, which may, for example, see an institution being tasked with an in-depth systematic enquiry into the area in order to identify the extent and source of risks. The third component is to identify existing systems or risk governance already in place within relevant institutions to identify and respond to the problem in hand. The final component is the selection of scientific criteria for risk assessment which involves the adoption of key assumptions, conventions and procedural rules for assessing risks and may involve the development of initial plans for the 'roll-out' of these conventions across relevant institutions.

The vivid potentials of new technologies, then, become not only aspects of debate about new technologies, but also part of broader attempts to define and redefine the relationship between the military and science, in order to establish predominant visions for the 'ethical' and 'secure' development trajectories for such fields. These broader apprehensions about the evolving nexus between innovation and security have led to the emergence of new forms of expertise directed at the ethical assessment of both emergent weapon technologies, and technologies of 'misuse' concern. This has also involved the building of new institutions designed to assess, exploit and manage technology in the name of security. These have taken three key forms. The first has involved the expansion of the scope of legal and ethical oversight over military innovation (i.e. military ethics); the second has been the addition of security concerns to the ethical evaluation of civilian fields (New and Emerging Science and Technology Ethics [NEST]); and third, there have been more ambitious attempts to integrate military and civilian innovation ethics—as reflected in the US in the launch of new centres in fields such as Biotechnology[34] and Artificial Intelligence (AI).[35] Each of these areas of development raises distinct challenges in terms of collaboration and also tends to be associated with a broad range of societal apprehensions about the appropriate role of the military and broader security apparatus in technology development. Collectively, these drives to govern through ethics also form part of attempts to establish new visions for 'ethical' and 'secure' development trajectories for such fields.

Thinking about this issue in terms of technology assessment, then, allows us to get to grips with the broader norms which shape with way in which societies attempt to evaluate and govern emergent technologies. However, as with the innovator's dilemma, the innovation dilemma framing comes with its own shortcomings. The first, is that technology assessment is primarily a national-level practice; this means that assessments tend to reflect national-level assumptions about the appropriate purpose of innovation. Technology assessment practices may help to encourage sobriety about the threat and opportunities presented by technology in terms of national and international security. Furthermore, more democratic approaches may, of course, allow for resistance to militarising drives. However, they appear to reproduce more systemic forms of insecurity, and in particular, possessive rather than collective visions of international security. Finally, most approaches to technology assessment are dependent on the ability to exert influence on emerging fields. This is problematic, in that without international norms of collaboration in this area, control of emergent technology remains the reserve of the state developing it—despite its potential global down-stream effects. These issues are examined to a greater extent from perspectives adopted in the next section, which focuses on the inherent global security paradox of innovation.

The Global Insecurity Paradox

This paradox relates to the apprehension that innovation is a wellspring of new technology which is always developed somewhere, by someone for some purpose, but which can have unforeseen and potentially global effects. This essentially generates a type of principal agent problem. However, while it is increasingly apparent that the route to national security is a more global approach to the management of technology, state-centric conceptions of security are at the centre of current approaches to global security. These state-centric conceptions also tend to reflect the interests and values of established global powers and are associated with embedded power struggles between states. As with the other paradox in this book, this apprehension is old. However, our contemporary understandings reflect contemporary national-level approaches to framing technology as a source of military security, and of national security more broadly. Contemporary discussions of this issue have focused in particular on drives by the US, and to a lesser extent European states, under the auspices of the so-called third offset strategy. This makes sense, in that the US and European states remain global leaders in terms of the development of both civilian and military technology. However, it is clear that there is a need for more work which seeks to impartially examine how these issues are dealt with outside of these states. As made apparent most recently in discussions about lethal autonomous weapon systems, a key challenge is developing more collective conceptions of technology assessment and governance, as it is not only specific technologies, but whole systems of innovation, which are constructed as threats. For example, it is often argued that greater public transparency and accountability reduce the extent to which states will pursue 'immoral' technologies, and that within autocratic regimes, there are no such guarantees. Contemporary discussions of this issue also reflect the fact that the global institutions of both innovation and technology control have emerged primarily since the end of the Second World War. Both national and more globalist apprehensions have given rise to a series of apparent dilemmas of control, which are briefly unpacked before they are further contextualised. In each case, there is a different understanding of 'technology' as well as the impacts of technology state interests and relations, which lead to different approaches to management.

The Arms Limitation Problem

This apprehension centres on the idea that insecurity drives the development and stockpiling of weapons, but that the development and stockpiling of weapons drive further insecurity. Often referred to as the 'dollar auction' problem, states become locked into cycles of ever-increasing expenditure, without seeing a net benefit to their security. In response, states seek to find a way of reducing competition. This challenge then is framed in strategic terms, with a particular emphasis placed on material-linked threat perceptions, but a wide range of strategies focusing on different intervention points have developed.[36] Arms limitation may place emphasis on like-for-like reduction, as characterised by much of nuclear arms control during the Cold War. However, it may also involve other forms of conditionality, as is often seen in post-conflict situations. Such limitations may also extend to complete multi-lateral disarmament for some classes of weaponry, as seen in the areas of chemical and biological warfare.

The Arms Proliferation Problem

The development and trade of arms are understood as important to national and international security; however, states also see unrestricted trade, both horizontally and vertically, as a threat to security. In response, states seek to retain control over who can accesses weapon technologies—through not supplying weapons to adversaries, and encouraging others to do the same. There are a number of non-proliferation regimes which have emerged, which rely on a range of unilateral and multilateral systems designed to control the flow of weapons internationally. This includes regimes directed at types of weaponry in which development, possession and proliferation in any form are entirely prohibited. It includes regimes in which the right of possession is only acknowledged for some states. It also extends to weapons in which trade between allies is understood as acceptable, but is controlled for certain actors and purposes.

The Multi-Use Technology Problem

This dilemma focuses on the apprehension that technology (which includes physical technologies, as well tacit and explicit forms of knowledge) can have applications which are understood to both direct benefit and pose direct challenges to security. This is because technologies can be foreseeably utilised by a wide range of agents for a wide range of purposes. These purposes may be essential to the national interest; however, they can also aid the enemies of states, and pose a threat to global order. This issue requires controlling the flows of technologies (which include physical, digital, explicit as well as tacit forms of knowledge) between 'users' and between 'uses'.

The Ambiguous State Capacity Problem

States support high technology-based capacities which can be utilised for a wide range of purposes. Distinguishing and predicting the way in which states will use such capacities now and in the future is particularly important in the context of prohibited weapon systems such as biological and chemical weapons, and in the nuclear sector. However, the 'dual-use' nature of these capacities creates the possibility of both hiding clandestine projects and maintaining an ability to switch between peaceful and hostile work (often referred to as breakout potential). In response, states seek to increase transparency in order to reduce the risks of misperception and cheating. This has typically focused on domestic efforts to improve abilities to detect and attribute hostile use, and at multilateral processes directed at setting shared standards of behaviour, and systems to assure compliance—including the development of independent systems of verification.

The De-Anchored Capacity Problem

Globalisation is understood to bring great economic and social benefits to states; however, these same processes may reduce state control over innovation and technological resources. Industries and capacities developed in the civilian sector may be 'repurposed' by any number of actors, who may apply them for clandestine functions. These actors may also maintain ambiguous relationships with those industries performing these clandestine functions. This then, can create significant challenges in terms of profiling the risks posed by such capacities, which may be employed in deniable, or semi-deniable, ways by enemies for a wide range of purposes. An example would be 'botnets', which can be established and used by hacker-groups to launch cyber-attacks—with or without the tacit support of a state. A key focus of responses to this issue is to encourage states to enforce domestic regulation, and to deter against the use of such capacities for either criminal or hostile intents through both multilateral cooperation and doctrines focused on improving abilities to deter, attribute and respond to illicit behaviour. In addition, it is clear that states have an interest in preventing competition and collective neglect, from allowing science to be neglected as a collective resource, as well as a potential source of risk.

In summary, it is clear that we respond to these various manifestations of the international security paradox at a time in which our key national and international security policy tools appear increasingly outmoded by changes in technology, warfare and the international order. We appear condemned to develop new, and ever-more complex and abstract metrics to assess and monitor technology. Warfare appears to have maintained its essential brutal and disorientating characteristics, but transformed into something which appears increasingly boundless, deniable and ambiguous. Furthermore, the normative and institutional architecture of our existing global institutions creek in the context of the continued onslaught of world history. In this context, states and other actors have come to look to emergent techno-scientific fields not only to imagine and develop technologies and technological visions, but also to generate new forms of governance and oversight. Such initiatives have worked within established bounds, defined by assumed state interests, as well as institutionalised approaches to both decision-making and implementation. However, this has also been met with the rise of increasingly pre-emptive, global and humanitarian visions of security, which seek to challenge rather than work within institutionalised conceptions of security, in the context of more established and emergent weapon control systems. This has been reflected, for example, in the renewed attention given to developing mechanisms of technology assessment in the context of International Humanitarian Law. This includes, for example, renewed attention to requirements to examine the impact of advances in science and technology in the context of disarmament regimes covering chemical, biological, space-based, and environmental weapons as well as weapon systems covered by the United Nations Convention on Certain Conventional Weapons (CCW). It also includes renewed attention to how states can better fulfil obligations under article 36 of the 1977 Additional Protocol I of the Geneva Conventions that requires states to review new weapons, means and methods of warfare. This final framing provides a specific way of thinking about the dilemma of innovation—placing prevailing 'national' and 'international' security institutions as well as broader 'visions' of global governance at the centre of analysis.

This final perspective also has its own limits, as it places prevailing 'national' and 'international' security institutions and discourses at the centre of its analysis, which draws the eye away from developments emerging from outside the gaze of these institutions. National and international security, we are often reminded, are both an ambiguous and powerful symbol, which seems to suspend our critical faculties. To adopt an international security framing risks ignoring the way in which technology emerges as part of the security order—and emphasising instead its instrumentalisation within it. It ignores, then, the pluripotent nature of innovation, and how this is developed and delimited in emergent fields—something which is much more present in discussions of the innovator's and innovation dilemma.

Notes

1. This work has benefitted greatly from engagement from the ethical, philosophical and sociologial dimensions of these questions provided by: Brian Rappert, *Biotechnology, Security and the Search for Limits: An Inquiry into Research and Methods* (Basingstoke: Palgrave, 2007); Brian Rappert,

Experimental Secrets: International Security, Codes, and the Future of Research (New York: University Press of America, 2009); Seumas Miller and Michael J. Selgelid, 'Ethical and Philosophical Consideration of the Dual-Use Dilemma in the Biological Sciences', *Science and Engineering Ethics* 13, no. 4 (December 2007): 523–80, https://doi.org/10.1007/s11948-007-9043-4; and Seumas Miller, *Dual Use Science and Technology, Ethics and Weapons of Mass Destruction* (Dordrecht: Springer, 2018). For an introduction to this issue, and in particular, the methodological challenges it raises for researchers, see Brian Rappert, *Biotechnology, Security and the Search for Limits: An Inquiry into Research and Methods* (Basingstoke: Palgrave, 2007).

2. Robert K. Merton, *The Sociology of Science: Theoretical and Empirical Investigations* (Chicago: University of Chicago Press, 1973).
3. Gabriel Abend, *The Moral Background: An Inquiry into the History of Business Ethics* (Princeton: Princeton University Press, 2014).
4. John Brock II, *Why the United States Must Adopt Lethal Autonomous Weapon Systems EBook: United States Army Command, United States Army School of Advanced Military Studies: Amazon.Co.Uk: Kindle Store* (Fort Belvoir: Defence Technical Information Center, 2017).
5. Scott D. Sagan, *The Limits of Safety: Organizations, Accidents and Nuclear Weapons*, New Edition (Princeton, NJ: Princeton University Press, 1995).
6. Daniel Ellsberg, *The Doomsday Machine: Confessions of a Nuclear War Planner* (New York, NY: Bloomsbury, 2017).
7. Dan Saxon, *International Humanitarian Law and the Changing Technology of War* (Boston: Martinus Nijhoff Publishers, 2013).
8. Thomas I. Faith, *Behind the Gas Mask: The U.S. Chemical Warfare Service in War and Peace* (Urbana: University of Illinois Press, 2014), 61, https://muse.jhu.edu/book/35211.
9. Edward M. Spiers, 'Gas Disarmament in the 1920s: Hopes Confounded', *Journal of Strategic Studies* 29, no. 2 (1 April 2006): 281–300, https://doi.org/10.1080/01402390600585092.
10. J. B. S. Haldane, *Callincus a Deffence of Chemical Warfare* (New York: E. P. Dutton, 1925); For further discussion, see Ulf Schmidt, *Secret Science: A Century of Poison Warfare and Human Experiments* (New York, NY: Oxford University Press, 2015), 61–62.
11. Sebastian Balfour, *Deadly Embrace: Morocco and the Road to the Spanish Civil War* (Oxford: Oxford University Press, 2002).
12. Lina Grip and John Hart, 'The Use of Chemical Weapons in the 1935–1936 Italo-Ethiopian War', in *SIPRI Arms Control and Non-Proliferation Programme* (SIPRI, 2009), 2, https://www.sipri.org/sites/default/files/Italo-Ethiopian-war.pdf.
13. Lina Grip and John Hart, 3.

14. Deborah Haynes, '150 First World War Mustard Gas Bombs Found at Beauty Spot', *The Times*, 18 October 2017, https://www.thetimes.co.uk/article/150-first-world-war-mustard-gas-bombs-found-at-beauty-spot-ttz7z99qm.
15. Wanglai Gao, 'China's Battle with Abandoned Chemical Weapons', *The RUSI Journal* 162, no. 4 (4 July 2017): 8–16, https://doi.org/10.1080/03071847.2017.1378408.
16. J. B. S. Haldane, 'Science and Future Warfare', *Royal United Services Institution Journal* 82, no. 528 (1 November 1937): 713–28, https://doi.org/10.1080/03071843709427314.
17. See, for example, Robin Black, 'Development, Historical Use and Properties of Chemical Warfare Agents', in *Chemical Warfare Toxicology* (2016), 1–28, https://doi.org/10.1039/9781782622413.
18. James M. Armitage et al., 'Environmental Fate and Dietary Exposures of Humans to TCDD as a Result of the Spraying of Agent Orange in Upland Forests of Vietnam', *Science of the Total Environment* 506–507 (15 February 2015): 621–30, https://doi.org/10.1016/j.scitotenv.2014.11.026.
19. Peter Williams and David Wallace, *Unit 731: Japan's Secret Biological Warfare in World War II* (New York: Free Press, 1989); Sheldon H. Harris, *Factories of Death: Japanese Biological Warfare, 1932–1945, and the American Cover-Up* (New York: Psychology Press, 2002); and Jeanne Guillemin, *Hidden Atrocities: Japanese Germ Warfare and American Obstruction of Justice at the Tokyo Trial* (New York: Columbia University Press, 2017).
20. Mark Wheelis, Lajos Rozsa, and Malcolm Dando, *Deadly Cultures: Biological Weapons Since 1945* (Cambridge, MA: Harvard University Press, 2006).
21. W. Seth Carus, 'A Century of Biological-Weapons Programs (1915–2015): Reviewing the Evidence', *The Nonproliferation Review* 24, no. 1–2 (2 January 2017): 129–53, https://doi.org/10.1080/10736700.2017.1385765.
22. See, for example, Brian Balmer, 'Killing 'Without the Distressing Preliminaries': Scientists' Defence of the British Biological Warfare Programme', *Minerva* 40, no. 1 (1 March 2002): 57–75, https://doi.org/10.1023/A:1015009613250.
23. Lentzos Filippa, *Biological Threats in the 21st Century: The Politics, People, Science and Historical Roots* (London: World Scientific, 2016), sec. III.
24. Dr. Armin Krishnan, *Killer Robots: Legality and Ethicality of Autonomous Weapons* (Farnham: Ashgate, 2013).
25. David Collingridge, *The Social Control of Technology* (New York: Palgrave Macmillan, 1981), 11.

26. On these contrasts, see Wolfgang Liebert and C. Schmidt, 'Collingridge's Dilemma and Technoscience', *Poiesis & Praxis* 7, no. 1–2 (1 June 2010): 55–71, https://doi.org/10.1007/s10202-010-0078-2.
27. Sheila Jasanoff, *States of Knowledge: The Co-Production of Science and the Social Order* (London: Routledge, 2004).
28. Jasanoff, *Designs on Nature.*
29. Jesse Ellman, Lisa Samp, and Gabriel Coll, 'Assessing the Third Offset Strategy', A Report of the CSIS International Security Programme (CSIS, March 2017), https://csis-prod.s3.amazonaws.com/s3fs-public/publication/170302_Ellman_ThirdOffsetStrategySummary_Web.pdf?EXO1GwjFU22_Bkd5A.nx.fJXTKRDKbVR.
30. Stephanie Carvin and Michael John Williams, *Law, Science, Liberalism, and the American Way of Warfare* (Cambridge: Cambridge University Press, 2014).
31. Ortwin Renn, Andreas Klinke, and Marjolein Asselt, 'Coping with Complexity, Uncertainty and Ambiguity in Risk Governance: A Synthesis', *AMBIO* 40 (3 February 2011): 231–46, https://doi.org/10.1007/s13280-010-0134-0.
32. On this, see, for example, Kathleen M. Vogel, *Phantom Menace or Looming Danger? A New Framework for Assessing Bioweapons Threats* (Baltimore: Johns Hopkins University Press, 2012).
33. O. Renn, *Risk Governance: Coping with Uncertainty in a Complex World* (New York: Earthscan/James & James, 2008), 48–51.
34. See, for example, DARPA's Innovation in Biotechnology Project, https://www.darpa.mil/about-us/innovation-in-biotechnology.
35. See, for example, the US DOD's newly established Joint AI Centertific field of security concern.
36. Stuart Croft, *Strategies of Arms Control: A History and Typology* (Manchester, UK and New York: Manchester University Press, 1997).

CHAPTER 3

Synthetic Biology as a Techno-Scientific Field of Security Concern

Abstract In this chapter, there is an introduction to the field of synthetic biology as a field of security concern. This involves characterising this area of innovation as an emergent 'techno-scientific' field—characterised by hype, ambiguity, institution building and power struggles. In this chapter, it is argued that fields such as synthetic biology are not only subject to security politics, but also help shape national approaches to security. The field is characterised as a disciplinary paradigm, as an emergent academic community, as a national scientific project and finally, as a field of national and international security concern.

Keywords Synthetic biology · Science and Technology Studies · Techno-science · Imaginaries

The term 'synthetic biology' has a long history within the biological sciences.[1] However, since the early 2000s, the term has become synonymous with communities of researchers, a range of practical and epistemic aims, as well as specific institutions and foundational technologies. Both the history and scope of the field of synthetic biology are contested. This is in part because such labelling is ultimately political. It is also in part because the term is always used as short hand for something else. In the following sections, various dimensions of the field, as an area of both innovation and governance, are introduced.

B. Edwards, *Insecurity and Emerging Biotechnology*,
https://doi.org/10.1007/978-3-030-02188-7_3

Synthetic Biology as a Disciplinary Paradigm

The term 'synthetic biology' is commonly used to denote a specific scientific paradigm of biological innovation, and in particular, the applications of engineering principle to biology—something which has been reflected in vivid mission statements given at scientific conferences and public meetings, and in often less exuberant but more technically informative scientific papers.[2] However, not only have these visions evolved over the first decade of the field, they are also not necessarily visions to which all of those scientists working on 'synthetic biology' linked projects necessarily subscribe. When you attend synthetic biology conferences, you come across scientists from a wide variety of fields, operating in an increasing range of national contexts. This means that they have fundamentally different takes on the prospects and means through which biological organisms and processes can be understood and controlled, and distinct senses of the primary purposes and societal value of their work.[3] They also have varying views as to the viability and ethics of the grander visions of the field espoused by thought leaders in this space. The distinction between what falls within the scope of a techno-scientific field and what does not is essentially arbitrary from a technical perspective—and tends to be more pragmatic from a funding and research publication perspective. These boundaries, of course, are continually drawn and redrawn as the field matures. Generally speaking, a number of approaches have been taken to defining the field. In the early years of the field's development in US in particular, there was been a tendency to focus on two major projects, which were contrasted in almost every sense. This included the way in which they were funded (public versus private), the way in which they approached the task of grappling with biological complexity (bottom-up versus top-down), as well as the way in which they made sense of the underlying biology of the systems they were working with. In addition, as the field became more established in the US, however, and also spread to Europe and later more globally, different ways of defining the field would also emerge. By 2008, three main areas of work were discernible: DNA-based device construction, genome-driven cell engineering and protocell creation.[4] And by 2009, a review would note that this could also be supplemented by work on the creation of unnatural genetic codes and proteins (xenobiology) as well as synthetic microbial consortia.[5] The later approach focuses on how to create systems of engineered microbes which can perform more complex functions than single cultures.

Indeed, by 2014, UK funding councils were distinguishing between six sub-fields[6]:

- Metabolic engineering: Attaining new levels of complexity in modification of biosynthetic pathways for sustainable chemistry
- Regulatory circuits: Inserting well-characterised, modular, artificial networks to provide new functions in cells and organisms
- Orthogonal biosystems: Engineering cells to expand the genetic code to develop new information storage and processing capacity (xeno nucleic acids) and protein engineering
- Bio-nanoscience: Developing molecular-scale motors and other components for cell-based machines or cell-free devices to perform complex new tasks
- Minimal genomes: Identifying the smallest number of parts needed for life as a basis for engineering minimal cell factories for new functions
- Protocells: Using programmable chemical design to produce (semi-)synthetic cells.

These distinctions reflected the distinct national level disciplinary and institutional landscapes the field was emerging into. In the UK, for example, the establishment of centres of synthetic biology reflected an acknowledgement of the type of large interdisciplinary centres which had given rise to the field in the US. Naturally, this resulted in much disagreement in the UK over definitions of the field, for example, in relation to the distinction between synthetic biology and other established paradigms such as genetic engineering and systems biology.[7] In addition, from a more policy-orientated perspective, the field would also be carved up based on how it mapped against existing regulatory frameworks. For example, the sub-field of protocells, would be excluded from later European definitions of the field, at least partly for this reason.[8]

Synthetic Biology as a Vanguard Community

The term 'synthetic biology' is also used to denote the vanguard community[9] which came to prominence in the US in the early 2000s. This emergent collation of innovators and policy-makers sought to establish a brand and vision which would help them compete for the financial and human resources to establish a new field of innovation. The roots of this community can be traced through an interdisciplinary research team, which was established in 2001 at the Massachusetts Institute of Technology (MIT). This team included Drew Endy, a genetic engineer; Thomas Knight, a computer engineer and Randy Rettburg, who had worked in the computer industry. It can also be placed in the context of a broader history of developments in a wide range of scientific fields, and in particular, genetic engineering.[10]

A synthetic biology vanguard came to include a prominent group of scientists and technologists based in the US with backgrounds in molecular biology, chemistry, computer informatics and engineering from universities and institutes such as Harvard Medical School, MIT, the University of California at Berkeley, the J. Craig Venter Institute (JCVI), the California Institute of Technology, the Lawrence Berkeley National Laboratory and John Hopkins University. While there were ideological disagreements within this community, as well as rivalry, this vanguard sought to push the field forward not only in terms of the foundational technology but also to overcome existing institutional obstacles which they felt hindered contemporary biotechnology research. These early advocates, although not always in agreement, were vocal on a range of issues. This included the need for a new interdisciplinary engineering-driven approaches to life-science research and the need to overcome existing obstacles to innovation which they felt were reflected in the existing US approach to the funding and commercialising biotechnology.

Synthetic Biology as a National and International Project

The first major investment into the emergent synthetic biology community came in 2006, when the National Science Foundation provided 10 years of funding at $5mil/year for a Synthetic Biology Engineering Research Center (Synberc). This was based at the University of California, Berkeley. The EU was quick to follow, and financed 27 multi-national collaborative synthetic biology projects.[11] The UK rapidly became the second largest investor in the field, launching a number of 'inter-disciplinary' research centres and networks, as well as collaborative initiatives between the academic community and industry,[12] estimated to total over £62 million in the period between

2004 and 2012.[13] The early investment in the US would be followed up with substantial further investment. By the end of 2014, the US government had invested approximately $820 million, with Department of Defence (DOD) investment constituting about 67% of total investments in the field.[14] Throughout this period, private investment also contributed to the translation of capacities and developments built through this investment into industrial applications. By 2015, there were around 100 synthetic biology products already on the market or near to entering it.[15] Private investment in the field also continued to rise. To give some sense of scale, one review of the top 50 companies has claimed that $1.7 billion was invested in the field in 2017 alone.[16] In addition, a small proportion of investment into applications would come from 'crowd-sourcing' initiatives rather than more established public and private models of investment. This form of investment totalled nearly £600,000 between 2013 and 2015.[17]

The field also quickly internationalised with the establishment of international annual conferences, US and European funding for collaborative projects, as well as the establishment of a number of dedicated global research centres.[18]

Distinct national visions of the field of synthetic biology have emerged,[19] which are the product of distinct national innovation cultures, as well as of self-conscious drives among the field's advocates to distinguish the field in terms of the way in which scientific developments are translated to real-world applications.[20] The early developers of the field of synthetic biology understood the importance of vision building from the outset. And this would be reflected, for example, in their attempts to articulate not only a specific vision, but also to identify and problematise the broader institutional structures and drivers, which would likely impact upon the fields' development. In the US context, this was reflected most explicitly, for example, in the work by Rob Carlson and colleagues who would distinguish a number of scenarios for the field, which pointed to tensions inherent in the US approach to biotechnology funding and regulation, which served to caution against stove-piping the field into off-the-peg models of innovations drawn from other fields.[21]

In the early years of synthetic biology, competing visions of how this could be best achieved emerged as the field expanded. These reflected distinct experiences of the vanguard community on state-funded 'big science' projects, as well as work in the private biotechnology sector.[22] In addition, the influx of scientists who had experienced the Silicon Valley boom of the early 1990s replaced greater emphasis on the value of open-source approaches.[23] In this respect, then, the establishment of a vision of synthetic biology as a field of innovation was also closely tied to attempts to establish new visions of *how* to innovate in different national contexts.[24]

The drive to develop visions for industrialisation was also bound with questions about public acceptance and regulation. Within both the US and the UK, the way in which the vanguard community engaged in these issues reflected broader institutionalised national-level norms in the governance of Ethical, Legal and Social Issues (ELSI) associated with new fields. In the US context, this drew upon a tradition of scientific self-governance, in which scientists were understood to have a role in proactively establishing ethical and safety guidelines ahead of regulatory intervention. In the field of synthetic biology, the mobilisation of the image of Asilomar would served both a more inward facing and outward facing purpose. On the one hand, it formed community visions of the purposes and limits of ethical responsibility, and on the other, it helped reassure the public that the community was developing the field in a responsible way. It would also motivate drives to seek technical solutions to potential laboratory safety and security problems. And in particular, to risks associated with the environment release of GMOs. This included drives to develop 'kill switches' as well as more orthogonal biological systems.[25]

In the UK, the scientific community was also understood to play a role in the governance of the field, including in terms of pre-emptive risk governance. However, this role was more narrowly circumscribed by existing regulatory frameworks covering GMOs (Genetically Modified Organisms) and laboratory safety.[26] Furthermore, it is clear that such work was primarily framed by research funders in terms of the need to address public resistance to the field, which reflected broader norms in the governance of emerging biotechnology in the UK—informed in particular by experiences with GMO governance.[27]

Synthetic Biology as an Emergent Techno-Scientific Field of Security Concern

Early concerns about the potential that insights and technologies developed by the emergent synthetic biology community fed into a wide range of initiatives designed to assess, manage and communicate risks associated with the field. In this section, a general overview of the type of initiatives we have seen directed is provided. These initiatives are also placed in the context of the broader challenges to assessing the impacts of advances of technology on the risk posed by biological and chemical terrorism.

The development of the field of synthetic biology in the US and Europe was associated with initiatives within key research institutions, usually involving collaboration with ethicists and other forms of expert in the governance of emerging technologies. These drives reflected broader institutionalised norms of biotechnology governance in different national contexts. The initiatives helped identify and map relevant professional, legal and international principles of relevance to the identification and management of potential hazards associated with the field, including security concerns. They also involved proactive engagement with regulators and wider civil society on areas of concern. In addition, ethics engagement formed part of early community-building drives, as well as more internationalised training and professionalisation initiatives as the academic field became more established, which included an emphasis on managing safety and security risks. A key challenge to developing guidelines in relation managing the direct misuse potential or research, as well as the pre-emptive identification of possible security concerns raised by new technologies, was the low level of awareness of militarization and proliferation concerns among practitioners within the field.[28] Another challenge was the fact that conventions for how to define and assess certain types of security concerns emanating from civilian life science research were still emerging—in national contexts and at the international level.[29]

At the national level, there were dedicated assessments of the implications of developments associated with the field by expert bodies within the intelligence communities. These focused primarily on the threat posed by the use of infectious disease as weapon of mass destruction and terror. Many of these assessments involved closed processes, but others were more open- and involved engagement with academic and scientific communities.[30] In the US, there were concerns that advances in the field might make clandestine projects cheaper and easier to hide, and that advances might reduce the amount of technical specialism required to produce certain types of biological weapon. These drives were also associated with renewed attempts to address the challenges of assessing the implications of technology as part of intelligence assessment and technology assessment—an area in which states were coping with transformations in both the security environment and the scientific and industrial base. Generally speaking, technology assessment has involved assessing the extent to which new technologies and combinations of technology supersede existing technical routes of weapon development.[31] This has been coupled with intelligence-based assessments which discern the capabilities and intentions of known and potential adversaries. The key concern is that the increasing availability of powerful technologies, coupled with techniques described in the open scientific literature, might help states overcome the major barriers to re-producing infectious weapon agents which have been developed in historic programmes, or reducing the footprint of such programmes. However, the impact that such advances might have on either the likeliness of effectiveness of bioterrorism was contested to a greater extent by Western non-proliferation and intelligence experts.[32] In recent years, there have also been comparable discussions about the impacts of synthetic biology in terms of producing known toxins, as well as of aiding the discovery of new toxins based on natural templates.[33]

In terms of assessing the impacts of developments of potential concern, there are significant challenges posed by intelligence and technology-focused assessment. This included the fact that a key source of information for assessment is former state programmes, which were always secretive. While new information is constantly being unearthed, it is clear that public, and even intelligence community understandings, of these historic programmes will always be partial, especially in those programmes continue to which have passed out of living memory.[34] It is also clear that while there is certainly much to be learnt from studying these programmes, the world has changed a lot since many of them were running—in terms of both technology and the security environment. In addition, there are only a small number of previous biochemical terrorist incidents to extrapolate from. In relation to incidents involving toxins and pathogens, which have been a central preoccupation of concerns about the field of interest in this book, the actual number is subject to some debate, but since the end of the Second World War, there are perhaps less than 10 cases identified within the existing literature, and several of these are contested.[35] There has been a recent uptick in poisonings, and attempts to both buy and sell Ricin and other toxins—if we include that in our definition of bioterror. However, this perhaps points more to the effects of cultural tropes (Ricin formed part of well-known plotline in the series *Breaking Bad*) and the impact of global online marketplaces rather than advances in technology.

Indeed, there are a wide range of factors, both technological and social, which would affect decisions of groups or states to pursue and employ biological or chemical weapons.[36] Discussions of the field also come in the context of broader disagreements about how to assess the impacts of technological developments on the significant tacit knowledge barriers to development in particular.[37] What is clear, however, is that the assessment of emergent technologies needs to take into account a wide range of variables, requires a considerable range of expertise, and would need to be ongoing—in order to account for more incremental and side-on technological developments, as well as shifts in the broader security environment. This includes the availability of technologies which may open up older or forgotten routes to weapon development or involve the employment of technology as part of comparatively unsophisticated forms of delivery—something which has been referred to as the potential 'shock of the old'.[38]

The history of chemical and biological warfare also points to the chameleon character of this category of weapon, and in particular, the wide range of uses that poisons and diseases targeted at humans and their environment have been put.[39] This points to the idea that new niches may open up for weapon systems to serve a range of political purposes—including economic sabotage, and to gain media attention and prestige.[40] The use of an agent from the 'Novichok' family of nerve agents in Salisbury, England in 2018, makes for an excellent case-study of both the terrorizing and communicative potential of such weapons. Such concerns will continue to drive recalibration of threat assessment, as well as the revisiting of exiting approaches to deterring and responding to such attacks.

It is clear, then, that there are a wide range of scenarios of use. There are also a wide range of variables relevant to assessing probability and likely consequence—which also points to a large number of intervention points. As a consequence, states have tried to focus on identifying both scenarios of concern as well as technologies which would help overcome more cross-cutting bottle necks to the weaponisation of pathogens, toxins and other synthesised forms of toxic agent.[41]

Intelligence assessments have been supplemented with reviews by government advisory bodies and scientific organisations directed at the field, dealing with a range of more traditional laboratory safety and environmental concerns associated with biotechnology. Such reviews have also dealt with security concerns, which reflects broader trends in the governance of the life-sciences in the early 2000s.[42] A key drive in this respect has been attempts to develop clearer technical definitions of the scope of scientific research which might present misuse concerns, and to develop relevant governance frameworks. The US has led the way in this respect in terms of the domestic appraisal of domestic governance systems and in supporting international dialogue. In terms of domestic policy, there has been a drive to establish a clearer definition of the proliferation risks associated with biotechnological innovation. In 2003, the National Research Council produced a landmark study on the these issues; a key innovation within the report was the identification of seven classes[43] of experiment involving pathogens and toxins with significant potential to produce insights which might reduce barriers to biological weapon development. Work as part of this report process also laid the foundations for the establishment of an advisory board to help the government to develop a national approach on the issue. The National Science Advisory Board for Biosecurity (NSABB) further refined the scope of concern arguing that emphasis should be placed on '*Research that, based on current understanding, can be reasonably anticipated to provide knowledge, products, or technologies that could be directly misapplied by others to pose a threat to public health and safety, agricultural crops and other plants, animals, the environment, or materiel*'.[44] The NSABB also engaged in a number of assessments of specific experiments of acute dual-use concern involving *avian influenza* research, as well as in relation to the field of synthetic biology. These review processes would have impacts in as far as defining the ethical expectations upon the US scientific community. In addition, a wide range of comparable problem definitions were developed in various national contexts over the following decade.[45] They also fed into national-level bioethical assessments. This included, for example, The Presidential Commission for the Study of Bioethical Issues (PCBI), which is a US body that provides authoritative expert advice on the governance of emergent biotechnology. At the international level, the field of synthetic biology was understood to have a number of potential impacts across the global prohibition regime against biological and chemical weapons. At the heart of this regime are two disarmament treaty systems which were established in order to ensure against the development, transfer, stockpiling and use of poisons and disease as a means of warfare. The first, is the Biological and Toxin Weapons Convention (BTWC) (1972) and the second is the Chemical Weapons Convention (CWC) (1993). These treaties are each the product of their distinct negotiation histories—varying in terms of procedural mechanisms and implementation culture, approach to compliance management, as well as different institutional frameworks to manage relations between respective member states.[46] The CWC has a sizeable treaty organisation, employing a secretariat of about 500, much of which is needed to support its verification work, which involves periodic on-site inspection at relevant research and industrial facilities. This is in contrast to the BTWC which has no system of verification and no comparable treaty organisation—it is often noted by those who work on the convention that the annual cleaning budget of the UN in Geneva exceeds that of the Convention. These differences are also reflected in terms of the scope, organisation and understood function of science and technology review processes within these treaties. In both cases, S&T review is understood to involve not only the function of reasserting the comprehensive and future proof nature of the categorical prohibitions these treaties embody, but also a range of other objects established by theses treaties—including advances which could help prevent and mitigate the effects of chemical and biological attacks. These treaties also operate in the context of a range of other multilateral processes and agreements directed at export control harmonisation, as well as WMD terrorism.[47]

It is also clear that concerns about the misuse potential of biotechnology also extend beyond poisons and disease to other areas of potential hostile exploitation. A recent review conducted by the US National Academies of Sciences identified several additional areas of potential concern including:

Modifying the human microbiome Manipulating microorganisms that form part of the population living on and within humans—for example, to perturb normal microbiome functions or for other purposes.

Modifying the human immune system Manipulating aspects of the human immune system, for example, to up- or down-regulate how the immune system responds to a particular pathogen or to stimulate autoimmunity.

Modifying the human genome Creating changes to the human genome through addition, deletion, or modification of genes or through epigenetic changes that modify gene expression. A subset of this category is the modification of the human genome through *human gene drives*, the incorporation of certain types of genetic elements into the human genome that are designed to pass from parent to child during reproduction and that would spread a genetic change through the population over time.

This has been supplemented by concerns about environmental modification weapons, directed against plants and animals which could be used to irreversibly damage eco-systems, through wiping out, or modifying entire species of organism. Something which builds upon a long history of agricultural warfare and terrorism. This has been reflected most recently, for example, in discussion of so called 'gene-drives' which could hypothetically be used to push an inheritable genetic trait into wild populations or microorganisms, animals and plants.[48] In addition, while concerns in this area have so far focused on the threat posed by biology as a direct-effect weapon, there are also broader concerns about augmenting the biology of combatants,[49] which raise a host of ethical[50] and more philosophical[51] questions.

In the following chapters, these issues are examined in greater detail.

Notes

1. L. Campos, 'That Was the Synthetic Biology That Was', in *Synthetic Biology: The Technoscience and Its Societal Consequences*, ed. M. Schmidt et al. (Berlin: Springer, 2009).
2. D. Endy, 'Foundations for Engineering Biology', *Nature* 438, no. 7067 (2005): 449–53.
3. See for example Evelyn Fox Keller, 'What Does Synthetic Biology Have to Do with Biology?', *BioSocieties* 4, no. 2 (1 September 2009): 291–302, https://doi.org/10.1017/S1745855209990123.

4. Maureen A. O'Malley et al., 'Knowledge-Making Distinctions in Synthetic Biology', *BioEssays* 30, no. 1 (2008): 57–65, https://doi.org/10.1002/bies.20664.
5. Lam, Carolyn M. C., Miguel Godinho, and Vítor A. P. Martins. "An Introduction to Synthetic Biology." in *Synthetic Biology: The Technoscience and Its Societal Consequences*, ed. M. Schmidt, A. Kelle, A. Ganguli-Mitra, and H. de Vriend (Berlin: Springer 2009), 23–48.
6. BBSRC, 'Synthetic Biology ERA-NET—1st Joint Call', Accessed 15 December 2014, http://www.bbsrc.ac.uk/funding/opportunities/2013/era-synbio-call1.aspx.
7. J. Y. Zhang, C. Marris, and N. Rose, 'The Transnational Governance of Synthetic Biology Scientific Uncertainty, Cross-Borderness and the "Art" of Governance', 2011, 6, http://royalsociety.org/uploadedFiles/Royal_Society/Policy_and_Influence/2011-05-20_RS_BIOS_Transnational_Governance.pdf.
8. Stefanie B. Seitz and Kristin Hagen, 'Inter- and Transdisciplinary Interfaces in Synthetic Biology', *NanoEthics* 10, no. 3 (1 December 2016): no. 1, https://doi.org/10.1007/s11569-016-0277-y.
9. Stephen Hilgartner, 'Capturing the Imaginary: Vanguards, Visions, and the Synthetic Biology Revolution', in *Science and Democracy: Making Knowledge and Making Power in the Biosciences and Beyond*, ed. Stephen Hilgartner, Clark Miller, and Rob Hagendijk (Oxford: Routledge, 2015).
10. Sebastian Pfotenhauer and Sheila Jasanoff, 'Panacea or Diagnosis? Imaginaries of Innovation and the "MIT Model" in Three Political Cultures', *Social Studies of Science* 47, no. 6 (1 December 2017): 783–810, https://doi.org/10.1177/0306312717706110.
11. OECD, *Emerging Policy Issues in Synthetic Biology* (London: OECD, 2014), 151–52, http://www.keepeek.com/Digital-Asset-Management/oecd/science-and-technology/emerging-policy-issues-in-synthetic-biology_9789264208421-en.
12. Synbicite Initiative website, http://synbicite.com/button-pages-section-header/applications/.
13. UK Synthetic Biology Roadmap Coordination Group, 'A Synthetic Biology Roadmap for the UK' (Swindon: Research Councils United Kingdom, 2012), 8, http://www.rcuk.ac.uk/RCUK-prod/assets/documents/publications/SyntheticBiologyRoadmap.pdf.
14. Synthetic Biology Project and Wilson Center, 'U.S. Trends in Synthetic Biology Research Funding', 2015, http://www.synbioproject.org/site/assets/files/1386/final_web_print_sept2015.pdf.
15. Synthetic Biology Project and Wilson Center.

16. Calvin Schmidt, 'These Fifty Synthetic Biology Companies Raised $1.7B in 2017', *SynBioBeta* (blog), 3 January 2018, 50, https://synbiobeta.com/news/fifty-synthetic-biology-companies-raised-1-7b-2017/.
17. Synthetic Biology Project and Wilson Center, 'U.S. Trends in Synthetic Biology Research Funding', 8.
18. OECD, 'The Bioeconomy to 2030: Designing the Policy Agenda', 2009, http://www.oecd.org/futures/long-termtechnologicalsocietalchallenges/42837897.pdf; OECD, *Emerging Policy Issues in Synthetic Biology*; Pfotenhauer and Jasanoff, 'Panacea or Diagnosis?'
19. See for example OECD, *Emerging Policy Issues in Synthetic Biology*.
20. Pfotenhauer and Jasanoff, 'Panacea or Diagnosis?'
21. Stephen Aldrich, James Newcomb, and Robert Carlson, 'Scenarios for the Future of Synthetic Biology', *Industrial Biotechnology* 4 (March 2008): 39–49, https://doi.org/10.1089/ind.2008.039.
22. Aldrich, Newcomb, and Carlson.
23. Arti Rai and James Boyle, 'Synthetic Biology: Caught between Property Rights, the Public Domain, and the Commons', *PLOS Biology* 5, no. 3 (13 March 2007): e58, https://doi.org/10.1371/journal.pbio.0050058.
24. Pablo Schyfter and Jane Calvert, 'Intentions, Expectations and Institutions: Engineering the Future of Synthetic Biology in the USA and the UK', *Science as Culture* 24, no. 4 (2 October 2015): 359–83, https://doi.org/10.1080/09505431.2015.1037827.
25. Markus Schmidt and Lei Pei, 'Improving Biocontainment with Synthetic Biology: Beyond Physical Containment', in *Hydrocarbon and Lipid Microbiology Protocols : Synthetic and Systems Biology—Tools*, ed. Terry J. McGenity, Kenneth N. Timmis, and Balbina Nogales, Springer Protocols Handbooks (Berlin, Heidelberg: Springer Berlin Heidelberg, 2016), 185–99, https://doi.org/10.1007/8623_2015_90.
26. Emma Frow, 'From "Experiments of Concern" to "Groups of Concern": Constructing and Containing Citizens in Synthetic Biology', *Science, Technology, & Human Values*, 25 October 2017, 0162243917735382, https://doi.org/10.1177/0162243917735382.
27. Brett Edwards and Alexander Kelle, 'A Life Scientist, an Engineer and a Social Scientist Walk into a Lab: Challenges of Dual-Use Engagement and Education in Synthetic Biology', *Medicine, Conflict and Survival* 28, no. 1 (2012): 5–18, https://doi.org/10.1080/13623699.2012.658659; Andrew S. Balmer et al., 'Taking Roles in Interdisciplinary Collaborations: Reflections on Working in Post-ELSI Spaces in the UK Synthetic Biology Community', *Science & Technology Studies*, 2015, https://sciencetechnologystudies.journal.fi/article/view/55340; and Claire Marris, 'The Construction of Imaginaries of the Public as a Threat to Synthetic Biology', *Science as Culture* 24, no. 1 (2 January 2015): 83–98, https://doi.org/10.1080/09505431.2014.986320.

28. Alexander Kelle, 'Synthetic Biology and Biosecurity. From Low Levels of Awareness to a Comprehensive Strategy', *EMBO Reports* 10 (August 2009): S23–27, https://doi.org/10.1038/embor.2009.119.
29. Anna Zmorzynska, et al., 'Unfinished Business: Efforts to Define Dual-Use Research of Bioterrorism Concern', *Biosecurity and Bioterrorism: Biodefense Strategy, Practice, and Science* 9, no. 4 (December 2011): 372–78, https://doi.org/10.1089/bsp.2011.0021.
30. See for example recent discussion of such processes by Kathleen M. Vogel and Michael A. Dennis, 'Tacit Knowledge, Secrecy, and Intelligence Assessments: STS Interventions by Two Participant Observers', *Science, Technology, & Human Values* 43, no. 5 (1 September 2018): 834–63, https://doi.org/10.1177/0162243918754673.
31. An example of one such rubric in the public literature developed for this purpose is Jonathan B. Tucker and Richard Danzig, *Innovation, Dual Use, and Security: Managing the Risks of Emerging Biological and Chemical Technologies* (London: MIT Press, 2012).
32. J. B. Tucker, 'Could Terrorists Exploit Synthetic Biology?', 2011.
33. Scientific Advisory Board, 'Convergence of Chemistry and Biology: Report of the Scientific Advisory Board's Temporary Working Group' (The Hague: OPCW, 27 June 2014), http://www.opcw.org/index.php?eID=dam_frontend_push&docID=17438.
34. Carus, 'A Century of Biological-Weapons Programs (1915–2015)'.
35. W. Seth Carus, 'The History of Biological Weapons Use: What We Know and What We Don't', *Health Security* 13, no. 4 (29 July 2015): 219–55, https://doi.org/10.1089/hs.2014.0092.
36. James Revill, 'Past as Prologue? The Risk of Adoption of Chemical and Biological Weapons by Non-State Actors in the EU', *European Journal of Risk Regulation* 8, no. 4 (December 2017): 626–42, https://doi.org/10.1017/err.2017.35; JP Zanders, 'Internal Dynamics of a Terrorist Entity Aquiring Biological and ChemIcal Weapons', in *Nuclear Terrorism: Countering the Threat*, ed. Brecht Volders and Tom Sauer (Oxford: Routledge, 2016).
37. Vogel, *Phantom Menace or Looming Danger?*
38. Revill, 'Past as Prologue?', 641.
39. Kai Ilchmann and James Revill, 'Chemical and Biological Weapons in the "New Wars"', *Science and Engineering Ethics* 20, no. 3 (September 2014): 753–67, https://doi.org/10.1007/s11948-013-9479-7.
40. Revill, 'Past as Prologue?', 630.
41. Perhaps the most sophisticated look at this in the public domain remains Tucker and Danzig, *Innovation, Dual Use, and Security.*
42. Caitríona McLeish and Paul Nightingale, 'Biosecurity, Bioterrorism and the Governance of Science: The Increasing Convergence of Science and

Security Policy', *Research Policy* 36, no. 10 (December 2007): 1635–654, https://doi.org/10.1016/j.respol.2007.10.003.

43. The report defined experiments of concern as those that: (1) would demonstrate how to render a vaccine ineffective; (2) would confer resistance to therapeutically useful antibiotics or antiviral agents; (3) would enhance the virulence of a pathogen or render a non-pathogen virulent; (4) would increase transmissibility of a pathogen—this would include enhancing transmission within or between species; (5) would alter the host range of a pathogen; (6) would enable the evasion of diagnostic/detection modalities; (7) would enable the weaponisation of a biological agent or toxin.
44. NSABB, 'NSABB Draft Guidance Documents', 2006, https://osp.od.nih.gov/wp-content/uploads/2013/12/NSABB%20Draft%20Guidance%20Documents.pdf.
45. Anna Zmorzynska et al., 'Unfinished Business: Efforts to Define Dual-Use Research of Bioterrorism Concern', *Biosecurity and Bioterrorism: Biodefense Strategy, Practice, and Science* 9, no. 4 (December 2011): 372–78, https://doi.org/10.1089/bsp.2011.0021.
46. See for example Ralf Trapp, 'Convergence at the Intersection of Chemistry and Biology- Implications for the Regime Prohibiting Chemical and Biological Weapons', Biochemical Security Project Paper Series (Bath: University of Bath, July 2014), http://biochemsec2030.org/policy-outputs/.
47. See Alexander Kelle, *Prohibiting Chemical and Biological Weapons: Multilateral Regimes and Their Evolution* (Boulder, CO: Lynne Rienner Publishers, 2014).
48. Kenneth A. Oye et al., 'Regulating Gene Drives', *Science* 345, no. 6197 (2014): 626–28.
49. David Malet, 'Captain America in International Relations: The Biotech Revolution in Military Affairs', *Defence Studies* 15, no. 4 (2 October 2015): 320–40, https://doi.org/10.1080/14702436.2015.1113665.
50. Maxwell J. Mehlman and Tracy Yeheng Li, 'Ethical, Legal, Social, and Policy Issues in the Use of Genomic Technology by the U.S. Military', *Journal of Law and the Biosciences* 1, no. 3 (1 September 2014): 244–80, https://doi.org/10.1093/jlb/lsu021.
51. Jai Galliott and Mianna Lotz, *Super Soldiers: The Ethical, Legal and Social Implications* (Oxford: Routledge, 2016).

CHAPTER 4

Synthetic Biology and the Dilemmas of Innovation

Abstract Innovation can produce both good and negative consequences. This then appears to generate conflicting ethical responsibilities for those that create and those that facilitate creation. Emergent communities of innovators seek to establish and enact ethical norms. A key challenge facing fields such as synthetic biology is that there are often competing visions both within and external to the community, which create challenges for early communities in terms of both developing and projecting these value systems. In this chapter, major initiatives in the field of synthetic biology are reviewed and critically assessed.

Keywords Asilomar · Responsible research and innovation · Biosecurity · Synthetic biology

The synthetic biology vanguard community took a leading role in the discussion of national security concerns associated with the field in the US, which would have fundamental impacts on discussions of the security implications of the field globally. Early interactions with US security agencies, and the discussion of security issues at scientific conferences, served to raise awareness of potential security concerns within the broader emergent community. In response to these concerns, a number of community-focused initiatives emerged. In this section, I focus in particular on how the first two major meetings of the community in the US

B. Edwards, *Insecurity and Emerging Biotechnology*,
https://doi.org/10.1007/978-3-030-02188-7_4

would help set the scope and terms of future discussions—not only in the US but also internationally.

Between 2002 and 2003, as initial groundwork for a new field was being laid, Drew Endy, a key field architect, was working on DNA synthesis at MIT. In the context of broader outreach by the US security community to scientists in the field, Endy was commissioned to chair a review of the security implications of the area by DARPA.[1] The process brought together around 50 scientists and other experts to examine the issue. Endy's report was not officially published, nor was it classified, and Endy shared information on the process widely within the broader community.[2] This process represented one of the earliest articulations of the potential security implications of the field, as well as the potential roles that the community could play in addressing these concerns.

Two issues dominated the discussion of the risks associated with the field in this early review. The first was the dual-use dilemma as had been defined in a recent major biosecurity review.[3] This dilemma revolved around the problem that: '*any useful technology might also be intentionally or accidentally misapplied to cause harm*'.[4] This was in addition to '*the probable inability to control the distribution of technologies needed to manipulate biological systems*'.[5] Endy framed the issue as centring on a race between the inevitable development of technological capabilities which might make the hostile use of biotechnology more likely and effective, and the challenge of developing technical solutions to manipulate biological agents and abilities to detect, respond to and mitigate these effects. It was noted that it '*seems obvious that future biological threats will increasingly arise* via *the intentional or accidental (i.e., "michanikogenic") application of biological technology. Importantly, the rate of biological threat emergence is likely to become great enough to overwhelm current response technologies. We are (appropriately) developing and deploying fixed assets against existing, relatively-static biological threats. However, future biological risks are likely to be greater in number, more sophisticated in design and scope, and more rapidly developed and deployed*'.[6] Endy also pointed to provisional criteria by which to assess the effectiveness of existing strategies to respond to these emergent threats:'*(1) How long does it take us (in days) to detect a new emerging infectious disease or engineered agent? (2) How long does it take us (in days) to understand how the agent works such that we can respond as needed? (3) How long does it take us (in days) to delivery of a response?*'.[7]

Clearly, then, the emergent US synthetic biology community was being framed as an essential aspect of the response to threats emerging from advances in biotechnology more broadly. At the same time, it was noted that '*the same technologies that are needed to help enable rapid responses to new biological threats could also be used to help construct the threats themselves. Thus, a strategy for addressing future biological risk must consider how future technologies can be best combined with non-technical solutions in order to minimize both the number of sources of future biological risks, and the scope of the risks themselves.*'[8]

This did appear to generate a requirement for oversight, but oversight which did not inhibit the ability of the US research community to respond to the threats created by the field which it (and other communities) would *inevitably* generate. This, then, from the perspective of the report, pointed to the need for community-centred initiatives which would mitigate some of the misuse concerns associated with the field while still allowing for the exploitation of the field. It was noted that 'biological engineering training could include professional development programmes and codes of ethics; a well-conceived and responsibly implemented plan for educating future generations of biological engineers would help to expand strategic human resources for future biological defense.'[9] In addition, it was suggested that this ethical community could play a role in supporting the development of screening standards within the DNA synthesis industry.[10]

However, different perspectives on how to deal with security concerns associated with the field, and in particular, gene-synthesis technology would emerge. There was early disagreement within the community about the appropriate roles and limits of self-governance. George Church, a leading scientist at the Harvard Medical School, led the development of an alternative vision of governance in relation to the issue of gene-synthesis technologies. As Church recalled in an interview:

> I had a close up view of the exponential technology and I just thought that security was subject to innovation as well as the underlying technology and that there was a moment an opportunity to innovate in terms of surveillance. Most people were talking about codes of ethics and that doesn't formally include surveillance [I thought] surveillance would be good if you want the sort of people that might do something accidentally or purposely hazardous should have reasonable expectation that some that they will be caught[11]

In early 2004, Church developed an alternative set of proposals, which he circulated widely.[12] This document considered the option '*of setting up a clearing house with oversight assigned to one or more of Homeland Security or the FBI*'. The document made two suggestions. First, it suggested that oligonucleotide (oligo) sequence[13] orders should be screened for similarity to select-agent pathogens. The paper also suggested that all use of reagents and oligos '*could be automatically tracked and accountable (as is done for nuclear regulations)*'. Church also discussed the potential for the de novo synthesis of select-agents by terrorists utilising current science and technology. This approach, then, which drew upon centralised material control traditions, sat at odds with the more open-source and decentralised visions of the field being developed.

These differing perspectives were reflected in discussions of the first synthetic biology conference held the following month.[14] Rob Carlson, a technologist and long-time collaborator with Endy, noted in a report on the meeting:

> Not for the first time in this circle did I hear suggestions of licensing for scientists and of strict controls on the distribution of technology and reagents. But such measures are not likely to be effective. Worse, they will instil a false sense of security.[15]

In this early period, then, there were no concrete responses, but these emergent perspectives set up the framing of a debate which would play out in the early years of the field. These early debates and initiatives would also serve to assert the central role of this community as experts and commentators in public discussions. This meant that in November 2005, when the *New Scientist* ran an expose on US industry screening practices—a response from the community appeared in a matter of days. Drew Endy asserted in an editorial piece which followed that his Lab would '*...only do business with companies that operate transparent procedures for screening gene-synthesis orders for potential bioweapons. If other researchers follow suit, rather than simply placing orders on the basis of cost or speed of delivery, the whole industry would be forced into adopting tougher standards*'.[16]

During 2006, Jay Keasling, a Professor of Chemistry at Berkley, led on a collaborative bid for a new Synthetic Biology research centre. The NSF was enthusiastic about the proposal; however, it notified Keasling that funding would be contingent on engagement with security concerns associated with the field.[17] In response, Stephen Maurer, a lawyer and academic also based at Berkeley was asked to join the group as ethics lead. Maurer had a long-standing interest in innovation, national security and community self-governance. Maurer's first task was to conceptualise how such engagement should occur—as requirements were only vaguely defined by the funders. His early work focused on the development of systems of monitoring as well as democratically negotiated community-wide standards.[18]

Maurer initially secured funding from the Carnegie Corporation of New York and MacArthur Foundation for a collaborative project which was designed to both study and facilitate global synthetic biology community action on issues of concern. The research project, involving interviews with experts, coordination with other institutions and working groups, and two public meetings, was designed to be the basis for community-based action, which would be voted on at the next major Synthetic Biology meeting (SB 2.0 to be held in 2006). A report produced a few months before S.B 2.0 suggested a set of policy options to be considered by the scientific community at the forthcoming conference asserted that:

> synthetic biologists share a deep understanding of the biosafety/biosecurity problem and – in some cases – emerging consensus about what can and should be done to manage it. Many options can be implemented through community self-governance without outside intervention.[19]

And furthermore that:

> Community self-governance provides a realistic and potentially powerful complement or alternative to regulation, legislation, treaties, and other interventions by outside entities.[20]

The report also confidently reassured those looking on that policy proposals were the result of consultation with the scientific communities through interviews which had set out to learn '*what members believe, want, and are prepared to vote for*' and that the policy proposals were derived '*from consent*'.[21]

It was intended that the vote would take place publicly at the SB 2.0 and involve scientists within the community. The resolutions within this document that directly addressed polynucleotide synthesis technologies included an insistence that the academic community only deal with gene-synthesis companies which had adopted best-practice screening procedures by a given date and on the engagement of scientists in research on screening to further facilitate the implementation of these best practices. The proposals also addressed research, appropriating the synthetic biology community as both experts in dual-use governance and implementers of security policy. The report states, for example, that

> Six years of almost continuous discussion have given synthetic biologists a solid understanding of biosafety/biosecurity risks and the available possible policy instruments for reducing them[22]

Various factors would conspire to undermine this initial experiment. Some factors were internal to the Synberc Project—members of the other research thrusts began private discussions which resulted in Maurer's proposal being pulled '*at the very last moment* by *Keasling and his colleagues*'.[23] Maurer suggests that there were several reasons that members of the organising group pulled the proposal for the agenda:

> Some needed more time to think about the ideas. Others were concerned that the conference needed a constitution before it could vote, or that a vote might be divisive. Some participants hesitated out of respect for the fierce opposition of activists[24]

These *activists*, headed by the ETC group, publicly criticised the 'Asilomar-style approach' suggested by Maurer's group. They argued that the Asilomar approach had been wrong when it was first adopted, and that it still reflected the wrong approach now. In an open letter circulated to conference attendees, they argued that there was a need for further engagement with global civil society before such decisions were made. They also argued that the evaluation of socio-economic, cultural, health and environmental impacts should extend beyond concerns of existing preoccupations with misuse, and that scientists alone could not hope to assess or control these impacts.[25]

The internal politics of this decision are also described in a recent book by Maurer examining the principles of self-governance initiatives. Maurer recalled that in the days leading up to the meeting, Drew Endy would '*suddenly demand that the vote be cancelled*'. And that:

> Keasling and several non-biologist policy professionals promptly called a meeting where they discussed and eventually agreed. While Endy never explained his reasons, other participants reported the absence of a 'constitution' authorizing voting, and the possibility that a vote would invite public attention, government scrutiny, and/or 'split the community'[26]

A further reason given for the Berkley project not achieving its original aims was that the initiative was trumped by the announcement of a Sloan Foundation-funded project involving the CJVI which would also discuss options for federal oversight. This was to be published after the SB 2.0.[27] Regardless of the causes, SB2.0 did not include a community-wide vote on the implementation of biosecurity actions. The incorporation of civil society groups such as the ETC at the SB3.0 and SB4.0 events essentially blocked the prospect of any such action in the future and the failure of the vote would reduce expectations upon conferences which followed.[28] However, despite the fact this project did not achieve its primary aims, this work did contribute to establishment of this community as a key factor in both discussions about and responses to security concerns, not only in the US but at the international level.

Indeed, in the early years, the academic field of synthetic biology was increasingly acknowledged as an emerging community with ethical sensibilities. As the field emerged, the desire to establish a legitimate space within which to develop security policy was initially demonstrated through work by the community leaders to encourage the DNA synthesis industry to develop and adopt security screening practices against customers and specific orders—something which contributed to the emergence of US and European standards, as well as the eventual development of federal guidelines in the US.[29] In this case, then, the voice of the community mattered, not only because of the growing public profile of leaders in the field, but also because this community represented a major customer to the industry.

In addition, these early drives became the basis for the emergence of new visions and practices of laboratory risk assessment and research ethics which would come to have global impacts. These were embodied in the annual student competition which was established under the auspices of the field. The main competition involves undergraduate and postgraduate students working over the summer at university laboratories to design and build new biological systems, using parts from the 'Biobricks' parts registry. This competition emerged initially as a US-based student competition, with the first competition involving 5 teams taking place in 2004. The competition grew and evolved over the following decade; by 2017, there were over 300 teams taking part from all over the world. The competition places an emphasis on both biosecurity and biosafety; Something which is reflected in the rules of the competition, the way in which project plans are reviewed centrally, and the prizes for innovation in this area. For example, the 2017 undergraduate Human Practices prize was awarded to a team working on directed evolution techniques, who developed a data input scanning tool designed to flag up potential safety and security concerns.

These early community initiatives embedded safety and security training into community building drives; something which also fed into attempts to globalise discussion of security impacts and into the field. There were a number of initiatives led by scientists and social scientists associated with the field which would feed into disarmament conferences. This included, for example, work conducted under the auspices of Synberc (Synthetic Biology Engineering Research Centre) and iGEM (International Genetically Engineered Machine competition)—which reflected broader drive to internationalise discussion. This included technical presentations at UN meetings[30]; in addition, a number of iGEM teams have been sponsored to attend disarmament conference meetings.

Scientists in this community also worked to draw attention to the need to support research on technical solutions to some of the security and safety challenges associated with research and industrial applications.[31] In addition, they helped shape the relationships which would emerge between ethics and scientific projects within the community, as well as the broader relationship between this community and the broader domestic regulatory environment.

Key Challenges

Early anxieties about security raised a number of different forms of concern for the synthetic biology scientific vanguard. Some felt that the public and policy-makers were unnecessarily concerned about the field and that the field required protection. Others felt that there was a need to reflect on the ways in which the field could best be exploited to serve the common good, which they felt required more proactive approaches to education within the community. They also sought greater involvement of broader society and regulators in making the decisions they faced in those early years. These responses can be situated in broader norms of scientific public engagement which have emerged since the later 1980s and more progressive visions of scientific citizenship engendered in contemporary approaches to scientific and business ethics. However, in addition to these types of apprehension, more abstract, more 'uncomfortable'[32] forms of concern would emerge, which could not be dealt with through these community-centred approaches.

These latter types of anxiety related to broader questions about whether certain concerns required a more radical departure from norms of risk governance, reflected, for example, in Church's proposal. A key characteristic of these early debates was a focus on two related, but distinct, issues. The first was a need to establish a community of practice, and to build collective norms of both research and risk assessment. The second was the need to navigate the broader political environment which surrounded the emergent fields.

A key challenge facing such endeavours is the fact that, in the early years in particular, activists both within and outside of the field are able to quite effectively mobilise public concerns. This is not only because the field is primarily communicated though ambitious vision statements, but also because these early debates often run ahead of official reviews, which can begin the process of 'thinning' the broad (and often ambiguous) range of concerns which surround the field. This means that not only do early visionaries often face push back from scientists who may have concerns, or more cynically competing professional interests, but also civil society groups, which can create significant press attention for from the community. A sense of vulnerability therefore often pervaded these early discussions; something which has a long heritage in the field of genetic engineering and contemporary techno-scientific fields more generally in both the US and Europe. In the US, these fears manifested primarily as concerns about public backlash based on security concerns about bioterror, and in Europe as concerns about public backlash over safety. As Claire Marris has argued based on her experiences as an Science and Technology Studies (STS) scholar working in the field in Europe,

> This synbiophobia-phobia has been the driving force.....Avoiding a repeat of 'what happened with GM' is routinely mentioned as a key objective, and

during these discussions, the shared assumption is that controversy is necessarily a bad thing.[33]

In addition, while there were successes in terms of educating younger scientists through the iGEM competition, there were also broader ongoing struggles within the community to define its goals over the next decade. This was reflected in early security debates about the field, and in broader discussions about the innovation model to be adopted for the field.[34] In the early years, the community defined itself against many of the organisational principles which had dominated previous large state-funded biotechnology programmes. The embedding of critical STS scholars in the project also ensured the articulation of equally radical visions of governance. However, more established visions and approaches to innovation have played an increasingly important role as the field has matured. This was reflected, for example, at a major community conference in Singapore in 2017,[35] which included presentations from scientists, social scientists, government funders and industry. The key sponsor of the event was Intrexon, a biotechnology company which was then chaired by Randal. C. Kirk, a noted industry visionary. The CEO spoke in a personal capacity at the meeting and gave a slick keynote address which focused on emerging applications under development in agriculture and health, and the challenges associated with these fields in terms of securing investment and navigating the global regulatory environment. His talk included the discussion of work which has continued to attract criticism from civil society groups.[36] As with the Synthetic Biology community, an aim of this talk was to distinguish Intrexon as market leader in terms of the organisation's approach to investment, product development, market entrance and in terms of how it engaged societal concerns. However, the more corporate ethos still sat at odds in some respects with that of the broader scientific community. The talk which was to follow from the conference chairs Drew Endy and Matthew Chang highlighted three themes.[37] The first was the need to develop new visions for the scientific and technological aims of the field in the wake of the substantial progress which had already been made. The second, was the requirement to think about how to address the challenges presented by the loss of biodiversity and other environmental challenges. The third, was to continue to build a global, collaborative and open community which would make the world a better place. Later, during the conference, Jane Calvert, a University of Edinburgh social scientist who has been working in synthetic biology for over a decade, argued that the field was closing as a creative space, and that the parameters of success are narrowing around what is commercially viable.[38] These issues, related to the aspirations of emergent techno-scientific communities to transform the broader context they operate, including the relationship between scientists and embedded social scientists, will be explored in greater depth in the following chapter.

Notes

1. This work built upon an earlier small study led by Tom Knight, a computer Engineer based at MIT, who would also become a leading figure within the field. There is further background available here: https://dspace.mit.edu/handle/1721.1/38455.
2. Drew Endy, 'Strategy for Biological Risk & Security', 2003, https://dspace.mit.edu/bitstream/handle/1721.1/30595/BioRisk.v2.pdf?sequence=1; Drew Endy, '2003 Synthetic Biology Study', 14 August 2007, Dspace@MIT, https://dspace.mit.edu/handle/1721.1/38455.
3. National Research Council, *Biotechnology Research in an Age of Terrorism* (Washington, DC: The National Academies Press, 2004).
4. Endy, 'Strategy for Biological Risk & Security', 2.
5. Endy, 2.
6. Endy, 3.
7. Endy, 'Strategy for Biological Risk & Security'; Drew Endy, '2003 Synthetic Biology Study'.
8. Endy, 'Strategy for Biological Risk & Security', 5.
9. Endy, 5.
10. Endy, 5.
11. Interview with author, 2011.
12. G. Church, *A Synthetic Biohazard Non-Proliferation Proposal* (May, 2004).
13. Oligos are short chains of single stranded DNA molecules (or RNA) which are short (less than 200bp) and have a range of applications within research.
14. Held in Cambridge, MA 2004.
15. Robert Carlson, 'Synthetic Biology 1.0', *Future Brief*, 2005, http://www.futurebrief.com/robertcarlsonbio001.asp.
16. New Scientist, 'Editorial: The Peril of Genes for Sale|New Scientist', 9 November 2005, https://www.newscientist.com/article/mg18825252-000-editorial-the-peril-of-genes-for-sale/.

17. Paul Rabinow and Gaymon Bennett, *Designing Human Practices: An Experiment With Synthetic Biology* (Chicago: University of Chicago Press, 2012), 15.
18. Rabinow and Bennett, 16.
19. S. M. Maurer, K. V. Lucas, and S. Terrell, 'From Understanding to Action: Community-Based Options for Improving Safety and Security in Synthetic Biology', *University of California, Berkeley. Draft* 1 (2006): 1.
20. Maurer, Lucas, and Terrell, 4.
21. Maurer, Lucas, and Terrell, 5.
22. Maurer, Lucas, and Terrell, 25.
23. Rabinow and Bennett, *Designing Human Practices*, 18.
24. S. M. Maurer and L. Zoloth, 'Synthesizing Biosecurity', *Bulletin of the Atomic Scientists*, 2007, 18.
25. ETC Group, 'Backgrounder: Open Letter on Synthetic Biology', ETC Group, 23 May 2006, http://www.etcgroup.org/content/backgrounder-open-letter-synthetic-biology.
26. Stephen M. Maurer, *Self-Governance in Science: Community-Based Strategies for Managing Dangerous Knowledge* (Cambridge: Cambridge University Press, 2017), 116.
27. Stephen M. Maurer, 'End of the Beginning or Beginning of the End? Synthetic Biology's Stalled Security Agenda and the Prospects for Restarting It', *Valparaiso University Law Review* 45, no. 4 (19 September 2011): 1198–199.
28. Maurer, *Self-Governance in Science*, sec. 5.3.
29. See, for example, Maurer, 'End of the Beginning or Beginning of the End?'; 'Screening Framework Guidance for Providers of Synthetic Double-Stranded DNA', *Biotechnology Law Report* 30 (April 2011): 243–57, https://doi.org/10.1089/blr.2011.9969.
30. Kenneth A. Oye, 'On Regulating Gene Drives: A New Technology for Engineering Populations in the Wild' (6 August 2014).
31. Gautam Mukunda, Kenneth A. Oye, and Scott C. Mohr, 'What Rough Beast? Synthetic Biology, Uncertainty, and the Future of Biosecurity', *Politics and the Life Sciences* 28, no. 2 (2009): 2–26.
32. Sam Weiss Evans and Megan J. Palmer, 'Anomaly Handling and the Politics of Gene Drives', *Journal of Responsible Innovation*, 2017, 1–20; Claire Marris, Catherine Jefferson, and Filippa Lentzos, 'Negotiating the Dynamics of Uncomfortable Knowledge: The Case of Dual Use and Synthetic Biology', *Biosocieties* 9, no. 4 (November 2014): 393–420, https://doi.org/10.1057/biosoc.2014.32.
33. Claire Marris, 'The Construction of Imaginaries of the Public as a Threat to Synthetic Biology', *Science as Culture* 24, no. 1 (2 January 2015): 83–98, https://doi.org/10.1080/09505431.2014.986320.

34. Stephen Aldrich, James Newcomb, and Robert Carlson, 'Scenarios for the Future of Synthetic Biology', *Industrial Biotechnology* 4 (March 2008): 39–49, https://doi.org/10.1089/ind.2008.039.
35. SB 7.0 held the 13 June, 2017–16 June, 2017 at the National University of Singapore, Singapore.
36. See for example: http://www.etcgroup.org/issues/synthetic-biology.
37. Drew Endy and Matthew Chang, 'SB 7.0 Program Welcome & Charge' (13 June 2017), https://vimeopro.com/vcube/synbiobetasingapore/download/221375173.
38. Jane Calvert, 'Session 3: Art, Critique, Design and Our World' (14 June 2017), https://vimeopro.com/vcube/synbiobetasingapore.

CHAPTER 5

Synthetic Biology and the Dilemmas of Innovation Governance

Abstract Societies seek security through the development and maintenance of innovation systems, but innovation can also generate insecurity. This then appears to create conflicting demands for exploitation and precaution. In different national contexts, different approaches to managing this apparent dilemma have emerged. A key challenge raised by fields such as synthetic biology is that they tend to raise questions which go beyond the limits of existing expert knowledge and risk management systems. In this chapter, major initiatives to pre-emptively engage with such concerns are reviewed and critically assessed.

Keywords New and emerging science and technology · Imaginaries · Pre-emptive security governance · Disarmament

Early drives to establish safety and security practices were situated within and on the periphery of a vanguard of the academic community. In this chapter, these early initiatives are put into the context of broader attempts to establish and develop incremental as well as more radical promissory visions of the field as a national and international project. These visions related not only to what the field might eventually produce in terms of potential risks and benefits, but also more fundamental claims about how the field would develop. In particular, these visions involved promissory claims about both the process of innovation to be pursued by the field, and the process of societal assessment.[1]

B. Edwards, *Insecurity and Emerging Biotechnology*,
https://doi.org/10.1007/978-3-030-02188-7_5

This then constituted the emergence of responses to resolve the dilemma of control, discussed initially by Collingridge, in the context of a contemporary techno-scientific field.

In this section, there is an examination of the key initiatives which were central to the assessment of security implications of the field, which would emerge in the US and the UK. In both cases, these drives initially focused primarily on the academic research community—as a legitimate means to both identify and respond to anticipatory concerns which went beyond the scope of existing risk management systems. This role came to be reflected in how the field was defined in national-level policy discourse. In the US, this was reflected, for example, in the recommendations of the PCBI, and in the UK, in the innovation strategy which would be developed for the field.[2]

In the early years of the academic community, there would be sustained attempts to develop an expert consensus on the implications that advances in this field posed to US national security—and in particular in relation to concerns about bioterrorism. The promissory claims of developers within the US and the exuberant coverage this attracted within the media combined to generate a broad range of anxieties about the field. Some of these concerns related to specific ongoing projects, but at the same time, synthetic biology was also rapidly taking on a more emblematic significance in the context of broader ongoing debates within the US about the oversight of biotechnology. In the early 2000s, there were concerns expressed by biological and chemical weapon experts about a broad range of potential bioterror scenarios, as well as the implications of developments for the international control regimes directed at biological and chemical weapons more broadly.[3] The scope of discussions reflected not only preoccupations with bioterrorism in the US, but broader ongoing political developments in the area of US biodefense, homeland security and laboratory safety more generally—as well as the emergence of the DURC (dual use research of concern) agenda.[4] In this context, a number of initiatives sought to develop expert consensus on the scope of concerns emanating from research in field. These centred on the interrelated tasks of establishing a predominant problem framing, delimiting an appropriate scope for the search for concerns stemming from the field, as well as the development and legitimation of new metrics of risks assessment.

Much of the early work would be supported by a $10 million grant from the Alfred. P Sloan Foundation, which began in 2005.[5] This initiative provided funding for a number of projects which developed a deeper technical understanding of the issues, and also supported continued community engagement in the field—including the amateur biology community, whose affiliation to the core academic community was cemented by Drew Endy's vision of the future of the field. In terms of developing deeper technical understanding of the security risks associated with the field, the Sloan Foundation also supported a number of early projects directed at assessing the societal implications of the field. The most significant assessment of the security implications of the field was led by the policy centre at the JCVI, which had been addressing concerns associated with its own major drive to develop 'artificial cell' since 1999.[6] The key output of this was the 2007 "Synthetic Genomics: Options for Governance" report. In terms of facilitating continued community engagement with security concerns, which necessitated input from a wide range of experts, the Foundation also supported the National Academies of Sciences to establish a Forum on Synthetic Biology in 2012 to facilitate discussions about scientific, technical, ethical, legal, regulatory, security, and other policy issues associated with the field, which involved a number of meetings in the US. This was in addition to supporting work at the Woodrow Wilson International Centre to: identify risks, evaluate the adequacy of existing regulatory mechanisms; and to educate policy-makers, journalists and the public about synthetic biology. The project also supported the discussion of the social implications of developments in the field at major scientific conferences in the field. The Sloan Project finished in 2014, but left legacy funding for the Synbio Leadership Excellence Accelerator Program (LEAP). This program continued to provide training to academic and industry scientists on ethical, legal and social issues.

In addition to this, engagement with social issues was also supported through the inclusion of ethics components in major research projects, most notably, in the NSF-funded project at Synberc. When the NSF invested into this project, it did so under the proviso that a certain proportion of the funding be spent to address the social implications of the work. This translated to the establishment of an ethics, ELSI or 'human practices' thrust, which engaged in a range of initiatives directed at laboratory safety and security and national-level policy development. This early support produced a body of literature, helped establish a loose community of experts, and fed into assessment initiatives in the US and internationally. Most notably, this included work under the auspices of the NSABB [7] as well the PCBI.[8]

Despite the US-centric nature of these early initiatives, they would also have fundamental effects in other states where comparable national synthetic biology projects were being established. In particular, this included building links between policy shapers in the US and UK. It also meant that security concerns which had emerged about the field in the US were discussed alongside more traditional safety and ethics concerns about genetic engineering in Europe. For example, misuse issues about the emergent field were raised in a 2005 EU high-level report and received the lion's share of commentary in the section of the document on potential risks associated with the field.[9] Security concerns were also reflected in an early European initiative under a 2-year EU-FP6-funded project called 'SynbioSafe', which examined the safety and ethical aspects of synthetic biology.[10]

Security concerns and early community security initiatives in the US were also reflected in early reviews of the ethical and social implications of the field in Europe.[11] The UK became the largest investor into the field in Europe in the early years, primarily in the form of cross-research council funding for seven interdisciplinary research networks. This investment also included funding for an embedded ELSI component within these projects. These became the basis of a range of initiatives designed to foster public and expert engagement with a range of issues. Security concern was a new addition to longer standing concerns about the safety, moral and environmental impacts of genetic engineering research.[12]

In the following section, there is discussion of the key challenges these initiatives faced, focusing in particular on attempts to build expert consensus on the scope of concerns, build new institutions directed at these concerns and build a broader vision for the field. It also explores the vanguard of scientists would become central to these processes.

Key Challenges

In the early years of synthetic biology, attempts engage with pre-assessment of the field—in particular the establishment of a framing of the issue, and the search and screening of potential concerns—fell primarily to a loose network of expert review processes, initially in the US, but a European expert discourse also emerged. A key challenge here was the interdisciplinary and contested nature of an amorphous and rapidly evolving area of innovation. The task of developing strategies for the ongoing review and assessment of emergent concerns also fell to these expert bodies.

In both Europe and the US, a key challenge that these communities faced was securing broader interest and support for the implementation of policy recommendations, which by design ran beyond the immediate interests of key government departments.

In the US, for example, the development of government guidance on industry screening lagged behind community-driven attempts by several years, meaning that it was the academic community and industry which led on the issue. Likewise, in relation to concerns about research, many of the recommendations by both the NSABB and PCBI were dependent on the uptake by government at the national level, which was a notoriously complex and slow process. The NSABB assessed the field of synthetic biology only on the assumption that its previous recommendations on dual-use research would eventually be implemented by the government. In relation to the PCBI, the Woodrow Wilson centre set up a score card to monitor the implementation of recommendations, which highlighted the wide range of bodies and processes involved in implementing recommendations.

Likewise in the UK, a key challenge, in the absence of a board comparable to the NSABB, was the encouragement of cross-government discussion on issues raised by the field—in a national context in which bioterror concerns associated with life science research were less of a concern for the security establishment.[13] For example, since 2005, there had been press attention given to concerns about mail-order DNA, which had initially emerged in relation to the emerging the gene-synthesis industry. In the summer of 2006, a journalist at The *Guardian* newspaper raised concerns about current customer and order screening in UK companies providing mail-order gene-synthesis services. In this investigation, journalists ordered a small section of DNA encoding for a small part of the smallpox virus genome. In response, the Department of Business Innovation and Skills hosted a cross-departmental meeting on the issue. The focus of the meeting and the report that followed was the immediate feasibility of misuse by small terrorist groups or short mail-order sequences. It was argued that while screening should be improved, the short-term risks posed by this route of acquisition were essentially negligible because of the technically demanding steps still required to construct a viable virus. However, it was noted that '*technologies will advance such that pathogenic organisms could be constructed or (more likely) be modified more easily*' and that '*key organisations [were] to alert Government if they become aware of any significant advances which might lead to major technological changes and thus to increases in risk*'.[14] This early discussion, then, asserted that the gene-synthesis industry merited vigilance at this stage, but that no further review needed to be taken. In response to these early concerns, the UK security establishment began to monitor advances in the emerging academic field of synthetic biology as a field which might reduce some of the barriers to abuse of gene-synthesis technologies upon which the UK biotechnology sector was increasingly dependent. However, there would be little interest in conducting technical assessments of the broader security implications of the field at such an early stage. There was also minimal interest from the Ministry of Defence (MOD), which delayed an early investigation into potential threats and opportunities associated with the field by several years.[15] In 2007, the HSE produced a small horizon scanning piece on the broader field of synthetic biology, but this did not address misuse issues. This then meant that in both the US and the UK, the academic community and ethical components of the US and European projects would perform a range of roles in the early years of the field, including the screening, pre-assessment and pubic communication of security risks.

In this context, then, the scientific community and other expert bodies focused on two key areas of boundary work, which were tightly interrelated. The first, was the delimitation of the scope of immediate concern for wider society. This involved the application of *anomaly handing rationales*[16] commonly associated with new technology, such as calls for the incremental modification of existing systems of control, the modification of the process and technological products of innovation, and the deferral of decisions until more information was available. The second challenge was institutionalising self-governance roles which were central to early promissory visions of governance by the field's vanguard, and emerging national-level strategies for the field. These challenges will be examined in more detail in the following section.

One of the first central areas of consensus to emerge within this community was the need to focus on more established technologies. The Sloan process focused on advances in gene-synthesis. The NSABB and PCBI processes which followed took a broader view on what constituted synthetic biology, which at least in part reflected the field's development during the intervening period. However, emphasis was also placed on areas underpinning the emergence of industrial applications at that point, with emphasis placed on the 'bottom-up' and 'top-down' genetic-engineering-focused approaches to engineering which dominated the US field.[17] This meant that the focus of early screening efforts was narrowed to a smaller sub-set of emerging synthetic biology technologies—which might be conceivably and immediately misapplied. For example, developments in xenobiology and synthetic microbial consortia were largely externalised. This focus on established rather than emergent technology systems also drew attention to the potentials of down-stream management of emergent technology.

These early strategies to defining the scope of concern, on one level, reflected the solidification of a specific problem framing of the issue of gene-synthesis technology, which pushed back against the idea that containment was either a viable or desirable strategy. On another level, they also reasserted a vision of technological intervention in which potential down-stream effects of established technologies could be managed pre-emptively by responsible 'users' and 'suppliers'. This, then, suggested that other areas of development in the field might later be managed in the same way, meaning that intervention *before* establishment was less important. This aligned with an understanding of research ethics, in which scientists were central to governing risks associated with the practice of research, but that down-stream technological outputs could be governed by other systems of risk management.

Another problem presented by the field was the impacts that developments in this field might have on the global security environment. A key aspect of this, for example, were concerns about state-level misuse of advances in the field; however, while state-level misuse concerns were widely acknowledged, these fell outside the scope of early key reviews. The Sloan report noted:

> ....we do not deal with state-sponsored research and development programs. No governance measure imposed by a national government will be effective at constraining that government's own activities if the responsible officials within that government choose to evade, ignore, or interpret their way around them. Moreover, no measure taken by researchers, firms, or other non-state entities operating within a government's jurisdiction can necessarily be relied on to resist pressure by that government. In the current international system, the only way to deal with abuses of national governments is through the actions of other governments, either collectively or individually. Such mechanisms are beyond the scope of this study.[18]

This then emphasised the idea that such concerns, although potentially worth further appraisal, should be dealt with in a separate process—and that these concerns were not enough to delay the exploitation of the field by this community. This was despite the fact that state-level misuse was understood by these experts to be much more technically feasible in the short term as compared to the threat posed by sub-state bioterrorism.

Another key aspect of the central consensus, which emerged in both the US and Europe, would be the centrality of existing models of GMO and governance to both defining and addressing the challenges raised by synthetic biology. For example, in the US and Europe, it was clear that existing regulation covering genetic engineering and paradigms could be extended to next-generation synthetic biology techniques in both the US and Europe—even if existing systems of organism categorisation might need to be reviewed in light of the increasing scope to transform the genetics and morphology of naturally occurring organisms.

This was a non-controversial point in a technical sense, considering the centrality of genetic engineering techniques in the most established aspects of the field. From a risk management perspective, the contested heritage of synthetic biology, as well as the contested visions of its industrialisation, were of little immediate consequence. However, while the public was being reassured that the field could be governed as a genetic engineering technology, it was also being told that the field was not simply the next step in genetic engineering by its leading advocates—something which generated an apparent tension in which incremental systems of risk management were held in contrast with more revolutionary framings of the field.

The ways in which these tensions emerged and dissolved in different national contexts would be shaped by broader national styles of biotechnology governance. These reflected distinct assumptions about the appropriate role of the public in discussion, disagreements over how risks should be managed and assessed locally—as well as broader debates about the future of national-level biotechnology innovation strategy more broadly.[19]

At one level, debates centered on the construction of the synthetic biology community as a monolithic 'ethical community' which could perform both risk management functions and act as a transformatory force. On the one hand, scientists were framed as capable of monitoring and managing developments—and also as a more reflexive and proactive community which could help transcend the limits of existing risk management systems. On the other hand, critiques would centre on the limits of 'community ethics' as a way in which to govern emergent fields. It was clear that while there was an awareness of security concerns within the community, experts, let alone bench scientists, were struggling to grapple with some of the anxieties raised by the field. At another level, critique focused on the broader logic of placing these communities at the centre of national discussions about technology assessment, when it was clear that this community only represented a transient intervention point in the governance of the field. As Maurer noted as early as 2007:

> Academic scientists still control the lion's share of synthetic biology projects, resources, and expertise. Potentially, this gives them important leverage over how industry evolves. But that will change. One company… is already using synthetic biology to make a parts-based organism. Other companies will surely follow, dwarfing the academic sector and eroding its capacity to influence events.[20]

In the early years then, it was not only the discussion of science which was subject to the effects of hype, but also the discussion of community self-governance. While the future of the field was open to community intervention in some respects, early community-centric visions had a tendency to underplay the broader institutional contexts in which the field would develop. It is also clear that there was disagreement on the broader political significance of engaging with pre-emptive risk assessment and that these issues reflected powerful institutionalised norms and values which ran much deeper than the academic field.

In the US, tensions were reflected in the often adversarial relationships between scientists and the Human Practices component at Synberc. This was reflected in the replacement of Maurer as the head of the Human Practices thrust in the wake of SB 2.0, and the turbulent period in which Rabinow led the thrust. Following concerns expressed by Keasling, the project PI, and eventually the NSF who were funding the centre, Rabinow was replaced by Drew Endy as the head of the Human Practices thrust and eventually resigned from his research positon at the centre. These disagreements reflected a long-standing dialectic between STS scholars and scientists on the political purposes and technical limits of pre-emptive risk governance in the context of emerging fields.[21]

Despite the replacement of Rabinow (initially by Drew Endy, and later by the political scientist Kenneth Oye), tensions continued. In 2015, for example, social scientists embedded at Synberc wrote a provocative article on ethical concerns associated with the engineering of a yeast strain which could produce opiates. The group argued that while such work had significant health benefits in terms of public health that the recent publication of a complete synthesis pathway now meant that '[i]n principle, anyone with access to the yeast strain and basic skills in fermentation would be able to grow morphine-producing yeast using a home-brew kit for beer-making' and that this could 'transform the illicit opiate marketplace to decentralized, localized production. In so doing, it could dramatically increase people's access to opiates.'[22] While the scientific community did not contest the revolutionary potentials of the field in terms of industrial production, there was public push back on the claim about the potential for home-brew style production. Following this incident, a senior synthetic biologist argued at a biosecurity event (under the condition of anonymity) that initial home-brew claims made by the social scientists had been designed to be provocative enough to guarantee publication in a top journal. Such remarks reflected broader tensions which have sometimes emerged between scientists and the social scientists embedded in their programmes.

In the UK, the relationships have appeared less tumultuous. However, they would still be associated with ongoing attempts to redefine the expected roles of ELSI practitioners.[23] In relation to security issues specifically, discussions have also reflected a situation in which the scope and content of ethical discussions about the field were skewed by early US preoccupations with bioterror concerns, rather than concerns emanating specifically from the UK field—something which appears to have contributed to the decision to push back against some of the 'myths' about the misuse potential of the field which have permeated European discussions of the field.[24]

In addition to these general tensions related to the broader politics of ELSI governance, and more local dynamics of research institutions, a major challenge as the field has matured from a vanguard community to a more institutionalised discipline has been the adaption of strategies of preemptive governance.[25] While this community may continue to remain a significant source of expertise as the field continues to spark industrialisation projects, it will not be able to continue to exhibit the same types of agency that it did in the early years of the field. The extent to which the principles, practices and concepts in these initiatives should or will transfer in practice in industrial biotech companies remains unclear. While it is apparent that there are tight links between many leading members of the field and many spin-off companies, it is also clear that there are different cultural norms around responsible innovation. One approach for the future might be to consider how existing approaches of industry align with RII principles developed for state-funded techno-scientific projects[26]; however, it is clear that industry may be informed by a much broader set of entrenched ethical norms.

In summary, then, the way in which synthetic biology emerged as a security problem was dominated in the early years by national quests to define both the scope of the concern and the character of the field as a site of innovation and governance. In both the US and UK, innovation was framed as a powerful, but controllable force which could be put to use in the national interest. Early governance drives involved practical community-centred risk management projects which asserted the centrality of synthetic biology research projects as hubs for innovation in terms of both technology and 'governance' at both the national and international level. In the following chapter, there is a deeper examination of the working assumptions which guided these early projects, which reveal the ways in which these initiatives would serve not only to help address, but also to reproduce more global security anxieties associated with the field.

Notes

1. Pablo Schyfter and Jane Calvert, 'Intentions, Expectations and Institutions: Engineering the Future of Synthetic Biology in the USA and the UK', *Science as Culture* 24, no. 4 (2 October 2015): 359–83, https://doi.org/10.1080/09505431.2015.1037827.
2. UK Synthetic Biology Roadmap Coordination Group, 'A Synthetic Biology Roadmap for the UK' (Research Councils United Kingdom, 2012), http://www.rcuk.ac.uk/RCUK-prod/assets/documents/publications/SyntheticBiologyRoadmap.pdf.
3. J. B. Tucker and R. A. Zilinskas, 'The Promise and Perils of Synthetic Biology', *New Atlantis* 12, no. 1 (2006): 25–45.
4. Gregory D. Koblentz, *Living Weapons: Biological Warfare and International Security* (Ithaca and London: Cornell University Press, 2009); Gregory D. Koblentz, 'From Biodefence to Biosecurity: The Obama Administration's Strategy for Countering Biological

Threats', *International Affairs* 88, no. 1 (2012): 131–48, https://doi.org/10.1111/j.1468-2346.2012.01061.x.

5. For a detailed review of the work of the Sloan foundation, see Gigi Kwik Gronvall, *Preparing for Bioterrorism: The Alfred P. Sloan Foundation's Leadership in Biosecurity* (Baltimore, MD: Center for Biosecurity of UPMC, 2012).
6. Gigi Kwik Gronvall, *Synthetic Biology: Safety, Security, and Promise* (Baltimore: CreateSpace Independent Publishing Platform, 2016), 110.
7. NSABB, 'Addressing Biosecurity Concerns Related to the Synthesis of Select Agents', December 2006, https://fas.org/biosecurity/resource/documents/NSABB%20guidelines%20synthetic%20bio.pdf; NSABB, 'Addressing Biosecurity Concerns Related to Synthetic Biology' (Washington, DC, 2010), http://osp.od.nih.gov/sites/default/files/resources/NSABB%20SynBio%20DRAFT%20Report-FINAL%20%282%29_6-7-10.pdf.
8. Presidential Commission for the Study of Bioethical Issues, 'New Directions: The Ethics of Synthetic Biology and Emerging Technologies', 2010, http://www.bioethics.gov/documents/synthetic-biology/PCSBI-Synthetic-Biology-Report-12.16.10.pdf.
9. NEST High-Level Expert Group, 'Synthetic Biology Applying Engineering to Biology: Report of a NEST High-Level Expert Group' (Luxembourg: European Commission, 2005), http://www.bsse.ethz.ch/bpl/publications/nestreport.pdf.
10. http://www.synbiosafe.eu/.
11. Most notably by the Rathenau Institute, an independent technology assessment organisation based in the Netherlands. Rinie van Est, Huib de Vriend, and Bart Walhout, 'Constructing Life-Early Social Reflections on the Emerging Field of Synthetic Biology' (Den Haag: Rathenau Institute, 2007), http://www.synbiosafe.eu/uploads/pdf/BAP_Synthetic_biology_nov2007%5B1%5D.pdf.
12. Brett Edwards and Alexander Kelle, 'A Life Scientist, an Engineer and a Social Scientist Walk into a Lab: Challenges of Dual-Use Engagement and Education in Synthetic Biology', *Medicine, Conflict and Survival* 28, no. 1 (2012): 5–18, https://doi.org/10.1080/13623699.2012.658659.
13. F. Corneliussen, 'Regulating Biorisks: Developing a Coherent Policy Logic (Part I)', *Biosecurity and Bioterrorism: Biodefense Strategy, Practice, and Science* 4, no. 2 (2006): 160–67; F. Lentzos, 'Rationality, Risk and Response: A Research Agenda for Biosecurity', *BioSocieties* 1, no. 4 (2006): 453–64; Filippa Lentzos and Nikolas Rose, 'Governing Insecurity: Contingency Planning, Protection, Resilience', *Economy and Society* 38, no. 2 (2009): 230, https://doi.org/10.1080/03085140902786611.

14. Department of Business Innovation and Skills, 'The Potential for Misuse of DNA Sequences (Oligonucleotides) and the Implications for Regulation', 2006, http://webarchive.nationalarchives.gov.uk/+/http://www.dius.gov.uk/partner_organisations/office_for_science/science_in_government/key_issues/DNA_sequences.
15. Confirmed by Email correspondence with the intended author report, on file with Author.
16. See, for example, Evans and Palmer, 'Anomaly Handling and the Politics of Gene Drives'.
17. This included work on Bio-fuels, Healthcare applications, Agricultural, Food, and Environmental Applications
18. M. S. Garfinkel et al., 'Synthetic Genomics: Options for Governance', *Industrial Biotechnology* 3, no. 4 (2007): 9.
19. Schyfter and Calvert, 'Intentions, Expectations and Institutions'.
20. Joachim Henkel and Stephen M. Maurer, 'The Economics of Synthetic Biology', *Molecular Systems Biology* 3 (5 June 2007): 117, https://doi.org/10.1038/msb4100161.
21. Rabinow and Bennett, *Designing Human Practices*; Anthony Stavrianakis, 'Flourishing and Discordance: On Two Modes of Human Science Engagement with Synthetic Biology' (Berkeley: University of California, 2012).
22. Kenneth A. Oye, J. Chappell H. Lawson, and Tania Bubela, 'Drugs: Regulate "home-Brew" Opiates', *Nature* 521, no. 7552 (18 May 2015): 281–83, https://doi.org/10.1038/521281a.
23. Andrew S. Balmer et al., 'Taking Roles in Interdisciplinary Collaborations: Reflections on Working in Post-ELSI Spaces in the UK Synthetic Biology Community', *Science & Technology Studies*, 2015, https://sciencetechnologystudies.journal.fi/article/view/55340.
24. Catherine Jefferson, Filippa Lentzos, and Claire Marris, 'Synthetic Biology and Biosecurity: Challenging the "Myths"', *Infectious Diseases* 2 (2014): 115, https://doi.org/10.3389/fpubh.2014.00115.
25. Gronvall, *Synthetic Biology*, 122.
26. Bernd Stahl et al., 'The Responsible Research and Innovation (RRI) Maturity Model: Linking Theory and Practice', *Sustainability* 9, no. 6 (16 June 2017): 1036, https://doi.org/10.3390/su9061036.

CHAPTER 6

Synthetic Biology and Dilemmas of Insecurity

Abstract It is increasingly apparent that the route to national security is a more global approach to the management of new and emerging technology. However, national security preoccupations still dominate in this area of global politics. This sets limits upon the types of concern associated with emergent technology which can be dealt with at the international level—particularly in the context of the broader UN System. Fields such as synthetic biology are significant in a number of ways at the international level—heralded as both a potential threat and benefit to global security. They are also a site at which more competitive and communal visions of security are reproduced. In this chapter, major initiatives to pre-emptively engage with security concerns are reviewed and critically assessed.

Keywords Disarmament · Biological weapons · Biosecurity

There are a number of key drives which have shaped the way in which the emergent techno-scientific field of synthetic biology has been understood to both constitute and contribute to broader security dilemmas at the global level. As noted from the outset of the book, the scope and character of the field, in a more technical sense, and in terms of societal implications, have been a continued source of contestation. This is particularly true when we examine the field from a more global perspective. It is clear that the field has increasingly come to be understood as

B. Edwards, *Insecurity and Emerging Biotechnology*,
https://doi.org/10.1007/978-3-030-02188-7_6

a complex web of scientific collaboration and industrialisation, nested in broader national projects. With this in mind, the discussion below is not intended to provide a comprehensive account of what the field has come to mean at the international level, in a more constitutional sense, but rather to trace some of the distinct contexts in which we can see more transnational attempts to develop a sense of areas of competition and collaboration. This provides a sense of the processes through which such fields come to be understood as areas of concern at the international level, and in particular, the way in which they can be understood to relate to broader developments at the international level.

A key driver of early security debates were community drives to establish a transnational academic community, which built upon existing innovation infrastructures. These drives opened up new areas of potential exploitation, and generated significant expectations about the economic and social benefits that advances would bring. Drives to establish the academic field were also tightly bound with inward-facing attempts to develop ethical practices within the emergent academic community—including the iGEM initiative. Aspects of this vanguard community engaged in more outward-facing attempts to shape the broader context of the field. A key dimension in the early years were attempts that gave the community a collective ethical agency—particularly in terms of shaping practices in the associated gene-synthesis industry. Later, once the academic field had matured, professionalisation drives would also engage with this issue. This included work by Synbio LEAP which focused on a broader set of practitioners in the wider economic and political ecosystem surrounding the emergent academic field.[1]

These community drives appeared alongside attempts to develop national-level visions for the field, which included attention to potential benefits that could be exploited from the field, as well as regulatory approaches. At the international level, the Organisation for Economic Co-operation and Development (OECD), as well as work in a number of other nation-specific and more global forums emphasised the distinct national approaches to innovation and regulation which were emerging. This work reasserted the specific advances associated with the field, as well as the broader institutional context into which the vanguard community was navigating.

This was supplemented by attempts to assess the potential impacts of these fields upon a number of global institutions. These drives came from within the vanguard community and core foundational institutions in the US and Europe. This included papers designed to stimulate broader discussion on the security challenges raised by the field, as well as the national experiences of addressing these issues—in addition to contributions which fed into more formal reviews of the impacts of S&T in the context of disarmament, and other treaty systems.

Indeed, attention was given to the impact of the emergent academic field and associated scientific advances, technologies and industry in the context of a number of UN treaty systems. This included, most prominently the area of disarmament, under the auspices BTWC and CWC. In the context of the BTWC, there was frequent reference to the field from 2006 onwards, but little progress in reaching a technical consensus on what the field actually was, or what challenges it actually raised—beyond the acknowledgement of its potential impacts across a number of treaty provisions. In contrast, in the context of the CWC, the SAB[2] as well as the SPIEZ laboratory in Switzerland[3] led more technical assessments which found there to be limited immediate impacts of developments in the field in areas such as verification. However, these assessments pointed to the need for ongoing monitoring as part of broader drives to better understand and respond to the impacts of increasing convergence between the biological and chemical sciences and associated industries.

There was also substantial work by scientific organisations which focused on supporting national implementation, as well as international-level reviews of scientific and technological advances. In the context of the CWC, this included contributions to technical reviews conducted by IUPAC.[4] In the context of the BTWC, included a major review of scientific trends conducted as part of a collaborative initiative between national scientific academies.[5] This work also fed into broader drives to help maintain and develop global ethical standards in science—particularly in the area of biosafety, biosecurity and dual-use research ethics.[6] There was discussion of the impacts of the field by biosafety professionals through organisations such as the Association for Biosafety and Biosecurity. Furthermore, such discussions would also be framed in the context of broader drives to improve public health responses to disease outbreaks as part of the global public health agenda.

These initiatives were supplemented broader work by the United Nations Interregional Crime and Justice Research Institute (UNICRI), which sought to develop a more holistic vision for the governance of security concerns raised by field. As a report on a series of expert meetings noted. Instead of only trying to control and deny access through international arms control measures, experts emphasized that the focus of securing biology should be shifted towards developing a shared responsibility between policy-makers'.[7] Such sentiments reflect broader transformations in the politics and practices of disarmament. This would be in addition to more speculative discussions of the impacts of the field on other treaty systems—including the Convention on Biological Diversity, which established a working group to examine the implications of the field.[8]

There were also reviews of the potential impacts of the development of the field in terms of the existing export control harmonisation systems. This included the Australia Group, which is a forum in which states maintain shared export control lists, and develop and share best practice.[9] The field was also reviewed by the UN1540 committee. An international body established under the auspices of United Nations Security Council Resolution 1540 in 2004, which obliges states to implement a domestic measures to prevent the proliferation of nuclear, chemical and biological weapons, and their means of delivery.[10]

The following section places these drives in a broader context, and reflects in particular on how the character of synthetic biology as a techno-scientific field has impacted upon its emergence as source of opportunity and concern.

Key Challenges

At the international level, the field of synthetic biology has been significant not only as a source of opportunities and challenges, understood to require acknowledgement and assessment—but has also formed part of broader processes of regime renewal, maintenance, and building. In the following section, some of the key ways in which the field has been understood as field requiring both control and exploitation are further unpacked. It focuses in particular on three types of challenge raised by the field: as a potential driver of arms racing behaviour, as a non-proliferation challenge and as a broader challenge to global systems of science governance.

From a more traditional arms control perspective, a key challenge raised by technological developments is the extent to which technological advances undermine existing sources of stability. This is in the sense that it makes cheating on agreements more tempting and difficult to spot. It is clear that synthetic biology may offer marginal advances to states in this regard; however, at the same time, this does not naturally equate to increased threat perceptions among states. This is in part because advances in foundational fields such as synthetic biology tend to bring benefits to defence; it is also because there are a range of other factors which shape how and why states choose to pursue lines of technological development for hostile purposes. The global prohibition against the development and use of poisons and disease not only severely curtails existing state-level development and exploration of the offensive military potentials in fields such as synthetic biology, but has fundamental impacts on how these fields are constructed as national projects, and in global diplomatic discourse. Indeed, inasmuch as fields of synthetic biology signify a potential challenge to existing systems of oversight and control, discussion and collective assessment of them have also helped to reassert the centrality of prohibitions in the future of politics in this area, and in particular, the future proof nature of categorical prohibitions. The buzz around field such as synthetic biology has also motivated collective attempts to review existing approaches to the implementation of these prohibition regimes at the global and international level. This then has helped to keep the door closed to the type of unfettered competition we saw between states in the Cold War era in this area—to the point where it is essentially unthinkable. It is worth remembering that such prohibitions not only protect the contemporary international order, but are also an essential part of it. This means that the gravest challenges to these prohibitions from this perspective then do not stem from new technology per se, but rather the erosion of faith in systems which have previously maintained that order, including unilateral deterrence, as well as the shifting, and increasingly covert character of warfare. Pre-emptive discussions of synthetic biology have tended to reflect, rather than underpin these concerns; it is one of a large number of areas of techno-scientific development which could be conceivably ripe for covert exploitation.

At the same time, where such categorical prohibitions do not exist in relation to the hostile exploitation of biotechnology, or they are more peripheral in contemporary international and domestic politics, it is clear that fields such as synthetic biology pose a more uneasy and ambiguous challenge in terms of the problem of like-for-like competition. For example, the potential relevance of advances in synthetic biology to the Convention on the Prohibition of Military or Any Other Hostile Use of Environmental Modification Techniques (ENMOD) has been alluded to since at least 2007.[11] This treaty prohibits the hostile use of environmental modification techniques having widespread, long-lasting or severe effects as the means of destruction, damage or injury to any other State Party. Recently, advances in gene-editing technology, specifically in relation to the potential of 'gene-drive' technology, have also led to some renewed interest in this convention, as well as other agreements covering GMO release, among these groups.[12]

A recent major study by David Malet notes the challenge facing those wishing to discern just how significant contemporary commitments to military exploration actually are in the area offensive biotechnology development: as well as the impacts exploratory drives have upon threat perceptions within other states. He does note, however, that in some cases, such exploration does seem to be contributing to cycles of insecurity. For example, expressions of interest by China in insects as vectors appear to have motivated both the US and India to invest in defensive work, which itself involves the mastery of some offense-related techniques.[13] The DARPA-funded Insect Allies programme, for example, seeks to develop counter-measures to naturally occurring or man-made disease outbreaks in crops—through the development of genetically engineered insects which can rapidly spread immunity from Specific diseases to crops.[14] It is also clear that current drives in areas such as human argumentation may raise similar competitive dynamics.[15]

It is, of course, clear that speculative proof-of-principle research is unlikely to be of pressing concern to competitors. However, perhaps a more significant impact such projects have is that they signal a closer working relationships between the academic research community and industry under the auspices of broader drives to renew and update the US military-academic-industrial complex. It is clear that there is an emergent range of configurations within these projects—in terms of ethical oversight and public accountability—which might also shape outside threat perceptions. This in an issue which merits much more substantial analysis—not only in the US, but also globally.

Another area in which techno-scientific fields such as synthetic biology raise challenges is non-proliferation. An area in which the physical and the digital are becoming increasingly interchangeable[16] and in which tacit barriers to the abuse appear to be under constant assault. The 'de-skilling' vision of synthetic biology in particular has been the most problematic, reflected in discussions of both the academic and affiliated amateur communities. Indeed, there were inherent tensions in relation to the extent and overall net benefits of more general de-skilling drives within early US discussions—with the US discourse coming to settle on the idea that the immediate risks were overplayed, and that the benefits of such an approach to the US national interests were significant economically and in terms of defence. It is apparent that as with other areas of technological innovation, the US continues to hedge its bets on the idea that it can always innovate it way out of binds, even ones of its own making.[17] It may well be right, it may be wrong; however it is clear that other states may have had different perceptions on the risk potentially raised by US industrialisation drives; for example, as a Chinese delegate would note at a BTWC meeting:

'The science and technology of synthetic biology are spreading rapidly and synthetic DNA technology has already become a basic tool of biological research; the related reagents and equipment are becoming ever easier to obtain. Accidental mistakes in biotech laboratories can place mankind in great danger. Synthetic biology in some civilian biotechnology research and applications may unintentionally give rise to new, highly hazardous man-made pathogens with unforeseeable consequences.'[18]

It is also clear that European states in particular have expressed more precautionary approaches in some areas, especially in relation to the release of GMOs. This then reasserts the importance of the establishment of shared principles in terms of both ethical content and process. And it is this which has become the focus of much discussion of the field of synthetic biology, as it has emerged in disarmament forums as well as others which focus on facilitating scientific and economic cooperation—designed to foster knowledge exchange and help with standard setting in areas such as safety, security and research ethics and, increasingly, precautionary approaches to technology assessment.[19] These principles are often understood to be under threat from the pressures of military, political and economic competition.

In summary, the field of synthetic biology has been framed as a challenge to existing global institutions; however, it has also reasserted the current and future importance of these institutions to national and collective security. Collective attempts to assess the field, where they have been forthcoming at the global level, reflect the limits of existing treaty systems—in terms of both scope and ambition. At the same time, discussion of this promissory has also underpinned calls for re-investment and renewal as part of a wider range of global governance agendas—linked to more progressive visions of global security. In this respect, the field has not only been changed as a challenge to the existing global order—but also as part of quests to build one. The conclusion regrounds these more heady observations and points towards more pragmatic steps which might be taken to manage the misuse of potential of biotechnology more broadly.

Notes

1. https://www.synbioleap.org/.
2. Scientific Advisory Board, 'Convergence of Chemistry and Biology: Report of the Scientific Advisory Board's Temporary Working Group' (The Hague: OPCW, 27 June 2014), http://www.opcw.org/index.php?eID=dam_frontend_push&docID=17438.
3. An up-to-date list of reports from this initiative is available here: https://www.labor-spiez.ch/en/rue/enruesc.htm.
4. Leiv K. Sydnes, 'IUPAC, OPCW, and the Chemical Weapons Convention', *CHEMISTRY International* 35 (2013), http://www.degruyter.com/view/j/ci.2013.35.issue-4/ci.2013.35.4.4/ci.2013.35.4.4.xml.
5. Katherine Bowman et al., *Trends in Science and Technology Relevant to the Biological and Toxin Weapons Convention: Summary of an International Workshop: October 31 to November 3, 2010, Beijing, China* (Washington, DC: National Academies Press, 2011), http://www.nap.edu/catalog/13113/trends-in-science-and-technology-relevant-to-the-biological-and-toxin-weapons-convention.
6. Dana Perkins et al., "The Culture of Biosafety, Biosecurity, and Responsible Conduct in the Life Sciences: A Comprehensive Literature Review," *Applied Biosafety*, 7 June 2018, 1535676018778538, https://doi.org/10.1177/1535676018778538.
7. UNICRI, 'Security Implications of Synthetic Biology and Nanobiotechnology: A Risk and Response Assessment of Advances in Biotechnology (Shortened Public Version)', 2012, ix.
8. AHTEG and Biology, 'Report of the Ad Hoc Technical Expert Group on Synthetic Biology', 7 October 2015, https://www.cbd.int/doc/meetings/synbio/synbioahteg-2015-01/official/synbioahteg-2015-01-03-en.pdf.
9. For example, in 2008, the Australia Group agreed to form a synthetic biology advisory body, https://australiagroup.net/en/agm_apr2008.html.
10. http://www.un.org/en/sc/1540/.
11. International Risk Governance Council, 'Guidelines for the Appropriate Risk Governance of Synthetic Biology', 2010, 25, https://www.irgc.org/IMG/pdf/irgc_SB_final_07jan_web.pdf.

12. Jim Thomas, 'The National Academies' Gene Drive Study Has Ignored Important and Obvious Issues', *The Guardian*, 9 June 2016, sec. Science, https://www.theguardian.com/science/political-science/2016/jun/09/the-national-academies-gene-drive-study-has-ignored-important-and-obvious-issues.
13. David Malet, *Biotechnology and International Security* (New York: Rowman & Littlefield, 2016), 171.
14. Todd Kuiken, 'DARPA's Synthetic Biology Initiatives Could Militarize the Environment', *Slate*, 3 May 2017, http://www.slate.com/articles/technology/future_tense/2017/05/what_happens_if_darpa_uses_synthetic_biology_to_manipulate_mother_nature.html?via=gdpr-consent.
15. Galliott and Lotz, *Super Soldiers*; Malet, 'Captain America in International Relations'.
16. Jean Peccoud et al., 'Cyberbiosecurity: From Naive Trust to Risk Awareness', *Trends in Biotechnology* 36, no. 1 (1 January 2018): 4–7, https://doi.org/10.1016/j.tibtech.2017.10.012.
17. Richard Danzig, 'Technology Roulette', May 2018, https://www.cnas.org/publications/reports/technology-roulette.
18. China, 'New Scientific and Technological Developments Relevant to the Convention: Background Information Document Submitted by the Implementation Support Unit', 2011, https://www.unog.ch/80256EDD006B8954/(httpAssets)/A72551B355472172C12579350035196F/$file/science_technology_annex.pdf.
19. Dana Perkins et al., 'The Culture of Biosafety, Biosecurity, and Responsible Conduct in the Life Sciences: A Comprehensive Literature Review', *Applied Biosafety*, 7 June 2018, 1535676018778538, https://doi.org/10.1177/1535676018778538; Jo L. Husbands, 'The Challenge of Framing for Efforts to Mitigate the Risks of "Dual Use" Research in the Life Sciences', *Futures*, 13 March 2018, https://doi.org/10.1016/j.futures.2018.03.007.

CHAPTER 7

Conclusion

Abstract This chapter draws together insights developed in the introductory and empirical chapters. It provides a summary of the distinct types of practical challenges that the field of synthetic biology has raised; as a source of ethical unease for scientific community, and as part of national and international quests to build and exploit and manage the field. A key argument made in this chapter is that the way in which the field has been addressed in these contexts has involved different preoccupations and working assumptions related to innovation and the prospects and limits of more globalist approaches to governance.

Keywords Technology assessment · Militarisation · Biological weapons · Techno-science

Over the past century, science, and the relationship between science, the state and the military—as well as the geography, planning and economics of innovation more broadly, have continued to transform. In this book, the challenges posed by contemporary configurations of science, national-level politics and international security have come into sharper relief. The relationship between science and security in the the context of emergent techno-scientific fields has been a key focus. This relationship has been explored through the examination of the emergence and establishment of the field of synthetic biology through three distinct perspectives. From each of these perspectives, the field of synthetic biology

B. Edwards, *Insecurity and Emerging Biotechnology*,
https://doi.org/10.1007/978-3-030-02188-7_7

has been understood to raise distinct types of challenge, which are now briefly summarised below.

The Innovator's Paradox

Innovation can produce both positive and negative consequences. This appears to generate conflicting ethical responsibilities for those that create and those that facilitate creation. This approach to thinking about the problem of misuse then places the ethical apprehensions of innovations and innovation communities at the centre of analysis. In particular, it emphasises the role of innovators as responsible actors, but also their roles in helping to define the extent and limits of that responsibility—in relation to the issues of misuse and militarisation. Techno-scientific fields such as synthetic biology raise a number of distinct challenges from this perspective; this includes their interdisciplinary nature, global character—as well as the new types of relationships such fields form with existing oversight systems and military funders. It is clear that the ethical apprehensions of scientists are an important focal point of public debate on such issues; it is also clear that innovators can play a wide range of roles in the governance of misuse concerns.

However, it is also clear that at the domestic level, the power of innovators is always to a great extent dictated to by prevailing innovation, regulative and economic norms. At the international level, it is apparent that the understood responsibilities of innovators vary—in relation, for example, to the extent to which they are expected to take activist roles. While then, there are many areas of consensus which stem from a desire to protect the global institution of science, scientists historically have also struggled to resist militarisation drives generated by more systemic sources of insecurity. As Julian Perry Robinson noted, now some 20 years ago, when discussing the history of scientist-led peace initiatives which emerged in the shadow of the bomb:

> Scientists have always had to contend with what has been called their "double loyalty": a sense of duty not only to their country but also to their science. In some disciplines, this duality could be conflictual, and, during those early years of the "national security state", it was indeed for more and more people. Loyalty to science is an abstraction not easily described or understandable outside its own world. In some scientists it is nonexistent, but in others it is passionate, overriding. Maybe it has to do with

> desire to protect newborn knowledge from deformation, from distraction, from loss, from waste. Maybe it also has to do with belief that science is for the common good.[1]

Such sentiments have motivated great progress in terms of the development of shared ethical standards globally. However, this ‘double-loyalty’ will always shape not only what scientists worry about, but what they don’t—which is largely determined by the broader institutional and cultural context they work within.

The Innovation Paradox

Societies seek security through the development and maintenance of innovation systems, but innovation can also generate insecurity. This appears to create conflicting requirements for both exploitation and precaution. This paradox revolves around two problems—how to plan innovation and how to manage it. In different national contexts, different approaches to answering these questions have emerged. The field of synthetic biology has raised a number of challenges. These relate not only to the inter-disciplinary character of the science, but also to conscious drives to distinguish the innovation and regulative strategies adopted within the field. In addition, the field would raise certain types of security concern which cut across existing forms of technical and more public facing expertise.

This approach to thinking about misuse concern emphasises the centrality of national-level projects to develop and assess such fields. Early discussion of synthetic biology reflected the specific national context in which it emerged—not only in terms of the scope of concern, but also the broader norms of assessment. It is clear that attempts to pre-emptively engage with discussion of potential proliferation and safety aspects at the international level may have helped to reassert the importance of safety standards, and relevant principles of global governance. However, a key challenge which remains is that national-level assessments potentially uncritically reproduce, rather than challenge, insecurities which are generated by state-level competition dynamics. This is because such assessments tend to be framed by national, rather than international interests. This then speaks to the challenges raised by the third paradox, set out below.

The Global Insecurity Paradox

It is increasingly apparent that global approaches to the management of innovation are important to both national and international security. The field of synthetic biology has made an interesting test case for regimes which already engage in technology assessment—particularly in the context of established disarmament and non-proliferation regimes. This has involved engagement (albeit in a more limited way) with the type of expert boundary-work seen at the domestic level—which has sought to delimit the impacts of the field of existing international institutions. An inherent limitation of this approach, however, is the extent to which existing institutional arrangements at the international level are prioritised in the discussion of the global challenges and opportunities presented by such fields. This then speaks to deeper questions about how to tackle inherent inequalities in terms of emergent approaches to global technology assessment. This is not only in terms of who is at the table, but also the values which should guide assessment. These questions are, of course, of fundamental importance and will also grow increasingly pertinent. This is because both the geography and economic of innovation is changing—in the context of shifting global power relations. Western states have grown accustomed to being the developers of technology; they need to become more comfortable with being the recipients.

Looking Forward...

The aim of this book was not to provide a plan for how such issues could be addressed but instead to sensitise the reader to the broader context in which these problems emerge. However, this being said, it is clear that in relation to biotechnology in particular, there are several types of initiative which might be of cross-cutting benefit to both assessing and responding to threats posed by the hostile exploitation of emergent technology—which as the very least merit further discussion. It is with these provocations that the book ends, motivated by the apprehension that the current pace of change in both technology and conflict suggests that it would be prudent to re-evaluate the collective principles which should delimit and guide military exploitation of biotechnology.

A key first step would be to identify areas and issues which merit particular attention from a more global perspective. This requires a clearer understanding of global investments into biotechnology by state militaries in particular. As it currently stands, there is no dedicated centralised and independent global data base on military investment into biotechnology. State militaries around the world invest into advanced biotechnologies in a host of ways, and for a host of purposes. There is a need to develop a clearer understanding of the types of purposes of such work. Without this, it is easy for publics to worry unduly, which can have negative impacts for benign work. Such knowledge gaps also allow doubts to fester, and for public fears to be played upon—which has historically helped states to justify both morally reprehensible and strategically misguided projects. Such work should focus not just on what is being invested in but also on the systems of oversight, transparency and public accountability in place. There is also a dire need for a re-examination and reassertion of the ethical limits which should be placed upon the military exploitation of biotechnology. This review must reflect the broad range of ethical concerns raised by contemporary biomilitary trajectories, which go far beyond the hostile exploitation of poisons and disease. This includes, but is not limited to, work on human and environmental modification. Forarmed with this knowledge, perhaps we can get a better grasp on the types of fronts which may open up in the future of disarmament.

Note

1. Julian Perry Robinson, 'Contribution of the Pugwash Movement to the International Regime Against Chemical and Biological Weapons' (1998), 5, http://www.sussex.ac.uk/Units/spru/hsp/documents/pugwash-hist.pdf.

Index

B. Edwards, *Insecurity and Emerging Biotechnology*,
https://doi.org/10.1007/978-3-030-02188-7

Lightning Source UK Ltd.
Milton Keynes UK
UKHW010134110219
337092UK00008B/151/P